Exploiting Linked Data and Knowledge Graphs in Large Organizations

Jeff Z. Pan · Guido Vetere
Jose Manuel Gomez-Perez
Honghan Wu
Editors

Exploiting Linked Data and Knowledge Graphs in Large Organizations

Editors
Jeff Z. Pan
University of Aberdeen
Aberdeen
UK

Jose Manuel Gomez-Perez
iSOCO Lab
Madrid
Spain

Guido Vetere
IBM Italia
Rome
Italy

Honghan Wu
University of Aberdeen
Aberdeen
UK

ISBN 978-3-319-83339-2 ISBN 978-3-319-45654-6 (eBook)
DOI 10.1007/978-3-319-45654-6

Softcover reprint of the hardcover 1st edition 2017

Printed on acid-free paper

This Springer imprint is published by Springer Nature
The registered company is Springer International Publishing AG
The registered company address is: Gewerbestrasse 11, 6330 Cham, Switzerland

Foreword

When I began my research career as a graduate student at Rensselaer Polytechnic Institute in 1989, the phrase "knowledge graph" was not in use. The use of graphs, however, as a notation for "knowledge representation" (KR) was quite common. CLASSIC, the first real implemented description logic, was just being introduced from Bell Labs, and although it had a linear syntax, the community was still in the habit of drawing graphs that depicted the knowledge that was being represented.

This habit traced its history at least as far as M. Ross Quilian's work on *Semantic Networks*, and subsequent researchers imagined knowledge to be intrinsic in the design of Artificial Intelligence (AI) systems, universally sketching the role of knowledge in a graphical form. By the late 1980s the community had more or less taken up the call for formalisation proposed by Bill Woods and later his student, Ron Brachman; graph formalisms were perhaps the central focus of AI at the time, and stayed that way for another decade.

Despite this attention and focus, by the time I moved from academia to industrial research at IBM's Watson Research Centre in 2002, the knowledge representation community had never really solved any problems other than our own. Knowledge representation and reasoning evolved, or perhaps devolved, into a form of mathematics, in which researchers posed difficult-to-solve puzzles that arose more from syntactic properties of various formalisms than consideration of anyone else's actual use cases. Even though we tended to use the words, "semantic" and "knowledge", there was nothing particularly semantic about any of it, and indeed the co-opting by the KR community of terms like semantics, ontology, epistemology, etc. to refer to our largely algorithmic work, rcliably confused the hell out of people who actually knew what those terms meant.

In my 12-year career at IBM, I found myself shifting with the times as a revolution was happening in AI. Many researchers roundly rejected the assumptions of the KR field, finding the focus on computation rather than data to be problematic. A new generation of data scientists who wanted to instrument and measure everything began to take over. I spent a lot of my time at IBM trying to convince others that the KR technology was useful, and even helping them use it. It was a losing battle, and like the field in general I began to become enamoured of the influential power of empirical evidence—it made me feel like a scientist. Still, however, my allegiance to the KR vision, that knowledge was intrinsic to the design of AI systems, could not be completely dispelled.

In 2007, a group of 12 researchers at IBM began working on a top secret moonshot project which we code-named “BlueJ”—building a natural language question answering system capable of the speed and accuracy necessary to achieve expert human-level performance on the TV quiz show, *Jeopardy!* It was the most compelling and interesting project I have ever worked on, and it gave me an opportunity to prove that knowledge—human created and curated knowledge—is a valuable tool. At the start of the project, Dave Ferrucci, the team leader, challenged us all to “make bets” on what we thought would work and commit to being measured on how well our bets impacted the ability to find the right answer as well as to *understand if the answer is correct*. I bet on KR, and for the first year, working alone on this particular bet, I failed, much as the KR community had failed more broadly to have any impact on any real problems other people had. But in the following year, Ferrucci agreed to put a few more people on it (partly because of my persuasive arguments, but mostly because he believed in the KR vision, too) and with the diversity of ideas and perspectives that naturally comes from having more people, we started to show impact. After our widely publicised and viewed victory over the two greatest *Jeopardy!* players in history, my team published the results of our experiments that demonstrated more than 10 % of Watson’s winning performance (again, in terms of both finding answers and determining if they were correct) came from represented knowledge.

Knowledge is not the destination

In order to make this contribution to IBM’s Watson, my team and I had to abandon our traditional notion of KR and adopt a new one, that I later came to call, “Knowledge is not the destination”. The abject failure of KR to have any measurable impact on anything up to that point in time was due, I claim, to a subtle shift in that research community, sometime in the 1980s, from knowledge representation and reasoning as an integral part of some larger system, to KR&R as the ultimate engine of AI. This is where we were when I came into the field, and this was tacit in how I approached AI when I was working in Digital Libraries, Web Systems, and my early efforts at IBM in natural language question answering.

The most ambitious KR&R activity before that time was Cyc, which prided itself on being able to conclude, “If you leave a snowman outside in the sun it will melt”. But Cyc could never possibly answer any of the myriad possible questions that might get asked about snowmen melting, because it would need a person to find the relevant Cyc micro-theory, look up the actual names and labels used in the axioms, type them in the correct and rather peculiar syntax, debug the reasoner and find the right set of heuristics that would make it give an answer, and even with all that it still probably could not answer a question like, “If your snowman starts to do *this*, turn on the air conditioner”, Watson might actually have had a shot at answering something like this, but only because it knew from large language corpora that 'snowman starts to melt' is a common n-gram, not because it understands thermodynamics.

Working with people from Cycorp, or with anyone in the KR&R world, we became so enamoured of our elegant logic that, without a doubt, the knowledge became our focus. We—and I can say this with total confidence—we absolutely believed that getting the right answer was a trivial matter as long as you had the knowledge and reasoning right. The knowledge was the point.

"Knowledge is not the destination" refers to the epiphany that I had while working on Watson. The knowledge was important, but it wasn't the point—the point was to get answers right and to have confidence in them. If knowledge could not help with this, then it really was useless. But what kind of knowledge would help? Axioms about all the most general possible things in the world? Näive physics? Expert Physics? Deep Aristotelean theories? No.

What mattered for Watson was having millions of simple "propositional" facts available at very high speed. Recognising entities by their names, knowing some basic type of information, knowing about very simple geospatial relationships like capitals and borders, where famous people were born and when, and much much more. Knowing all this was useful not because we looked up answers this way—*Jeopardy!* never asked about a person's age—but because these little facts could be stitched together with many other pieces of evidence from other sources to understand how confident we were in each answer.

This knowledge, a giant collection of subject-property-object triples, can be viewed as a graph. A very simple one, especially by KR&R standards, but this knowledge graph was not itself the goal of the project. The goal—the destination—of the project was winning *Jeopardy!* So, in fact, we made absolutely no effort to improve the knowledge we used from DBpedia and freebase. We needed to understand how well it worked for our problem in the general case, because there was no way to know what actual questions would be asked in the ultimate test in front of 50 million people.

Knowledge Graphs are Everywhere!

As of the publication of this book, most major IT companies—more accurately, most major information companies—including Bloomberg, NY Times, Microsoft, Facebook, Twitter and many more, have significant knowledge graphs like Watson did, and have invested in their curation. Not because any of these graphs is their business, but because using this knowledge helps them in their business.

After Watson I moved to Google Research, where freebase lives on in our own humongous knowledge graph. And while Google invests a lot in its curation and maintenance, Google's purpose is not to build the greatest and most comprehensive knowledge graph on Earth, but to make a search, email, youtube, personal assistants and all the rest of our Web-scale services, better. That's our destination.

Many believe that the success of this kind of simplistic, propositional, knowledge graph proves that the original KR&R vision was a misguided mistake, but an outspoken few have gone so far as to claim it was a 40+ year waste of some great minds. As much as I appreciate being described as a great mind, I prefer a different explanation: the work in KR for the past 40 years was not a waste of time, it was just the wrong place to start. It was solving a problem no one yet had, because no one had yet built systems that used this much explicit and declared knowledge.

Now, *knowledge graphs are everywhere*. Now industry is investing in the knowledge that drives their core systems. The editors of this volume, Jeff Pan, Guido Vetere, José Manuel Gómez Pérez and Honghan Wu, all themselves experts in this old yet burgeoning area of research, have gone to great lengths to put together research that matters today, in this world of large-scale graphs representing knowledge that makes a difference in the systems we use on the Web, on our phones, at work and at home.

The editorial team members have unique backgrounds, yet have worked together before, such as in the EU Marie Curie *K-Drive* project, and this book is a natural extension of their recent work on studying the properties of knowledge graphs. Jeff started at Manchester and has done a widely published work in formal reasoning systems, and moved to Aberdeen where his portfolio broadened considerably to include Machine Learning, large data analysis, and others, although he never strayed too far from practical reasoning, such as *approximate reasoning*, and querying for knowledge graphs. Guido has run several successful schema management projects on large data systems at IBM, and was part of the team that worked to bring Watson to Italy. Jose has done important research in the area of distributed systems, semantic data management and NLP, making knowledge easier to understand, access and consume by real users, and Honghan has been doing research in the area of medical knowledge systems.

After you finish this book, try to find a faded red copy of *Readings in Knowledge Representation* lest we forget and reinvent the Semantic Network.

May 2016

Dr. Christopher Welty
Google Research NYC

Preface

A few years after Google announced that their 'Knowledge Graph' would have allowed searching for *things, not strings*,[1] knowledge graphs start entering information retrieval, databases, Semantic Web, artificial intelligence, social media and enterprise information systems. But what exactly is Knowledge Graph? Where did it come from? What are the major differences between knowledge graphs for enterprise information management and those for Web search? What are the key components in a knowledge graph architecture? How can knowledge graphs help in enterprise information management? How can you build good quality knowledge graphs and utilise them to achieve your goals?

The main purpose of this book is to provide answers to these questions in a systematic way. Specifically, this book is for academic researchers, knowledge engineers and IT professionals who are interested in acquiring industrial experiences in using knowledge graphs for enterprises and large organisations. The book provides readers with an updated view on methods and technologies related to knowledge graphs, including illustrative corporate use cases.

In the last four years, we have been working hard and closely in the K-Drive—Knowledge Driven Data Exploitation—project (286348), which was funded by EU FP7/Marie Curie Industry-Academia Partnerships and Pathways schema/PEOPLE Work Programme. The main purpose of this project was to apply and extend advanced knowledge techniques to solve real-world problems, such as those in corporate knowledge management, healthcare and cultural heritage. Most of the challenges we encountered and techniques we dug into are highly related knowledge graph techniques. This book is a natural outcome of the K-Drive project that reflects and concludes the understanding we accumulated from the past four years of work, the lessons we have learned and the experiences we gained.

Contentwise, we will focus on the key technologies for constructing, understanding and consuming knowledge graphs, which constitute the three parts of this book, respectively. **Part I** introduces some background knowledge and technologies,

[1] Introducing the Knowledge Graph: things, not strings, googleblog.blogspot.com May 16, 2012

and then presents a simple architecture in order to help you to understand the main phases and tasks required during the lifecycle of knowledge graphs. **Part II** is the main technical part that starts with the state-of-the-art Knowledge Graph construction approaches, then focuses on exploration and exploitation techniques and finishes with advanced topics of Question Answering over/using knowledge graphs. Finally, **Part III** demonstrates successful stories of knowledge graph applications in Media Industry, Healthcare and Cultural Heritage; and ends with conclusions and future visions.

It is true that there is no *gold standard* definition of Knowledge Graph (KG). While working on the book, the editors and chapter contributors have debated lively on *what constitutes KG?*, *how is it related to relevant techniques like Semantic Web and Linked Data techniques?* and *what are its key features?* Fortunately, most, if not all, arguments have been settled and the conclusions and agreements have been put into the book, e.g. into the last two sections of Chap. 2. Even luckier, when finalising the book, editors have got the opportunity to collect opinions on *visions, barriers and next steps of Knowledge Graph* from key figures in the community including outstanding researchers, practitioners in leading organisations and start-ups, and representative users of various domains. Such valuable opinions have also been compiled into this book as part of its conclusion and future vision.

We would like to thank all of the chapter contributors as well as all members of the K-Drive project, who have given so much of their time and efforts for this book, in particular Dr. Yuting Zhao, who offered much helpful advice on the organisation of the book.

We had great pleasure in having Chris Welty write a touching Foreword for this book, sharing with us his rich experience and epiphany he had during the compelling BlueJ project, as well as his opinions on the motivation (‘*Knowledge Graphs are Everywhere!*’) and the importance of this book.

We would also like to acknowledge the IBM DeepQA research team for allowing us to use their architecture diagram marked as Fig. 7.1 in the book.

We are grateful to the following experts in the field for sharing with us their visions, barriers and next steps of Knowledge Graph in our concluding chapter: Sören Auer, Riccardo Bellazzi, Oscar Corcho, Richard Dobson, Junlan Feng, Aldo Gangemi, Alfio M. Gliozzo, Tom Heath, Juanzi Li, Peter Mika, Fabrizio Renzi, Marco Varone, Denny Vrandečić and Haofen Wang.

Aberdeen, UK
Rome, Italy
Madrid, Spain
Aberdeen, UK
June 2016

Jeff Z. Pan
Guido Vetere
Jose Manuel Gomez-Perez
Honghan Wu

Contents

Contributors

Panos Alexopoulos Expert System, Madrid, Spain

Ronald Denaux Expert System, Madrid, Spain

Alessandro Faraotti IBM Italia, Rome, Italy

Nuria Garcia-Santa Expert System, Madrid, Spain

Jose Manuel Gomez-Perez Expert System, Madrid, Spain

Marco Monti IBM Italia, Milan, Italy

Alessandro Moschitti University of Trento, Trento, Italy

Hai Nguyen University of Aberdeen, King's College, Aberdeen, UK

Massimo Nicosia University of Trento, Trento, Italy

Jeff Z. Pan University of Aberdeen, King's College, Aberdeen, UK

Fernanda Perego IBM Italia, Milan, Italy

Yuan Ren University of Aberdeen, King's College, Aberdeen, UK

Mariano Rodriguez-Muro IBM USA, Thomas J. Watson Research Center, Yorktown Heights, NY, USA

Kavitha Srinivas IBM USA, Thomas J. Watson Research Center, Yorktown Heights, NY, USA

Kateryna Tymoshenko Trento RISE, Povo di Trento, Trento, Italy

Guido Vetere IBM Italia, Rome, Italy

Boris Villazon-Terrazas Expert System, Madrid, Spain

Andrew Walker University of Aberdeen, King's College, Aberdeen, UK

Gemma Webster University of Aberdeen, King's College, Aberdeen, UK

Honghan Wu King's College London, London, UK

Yuting Zhao IBM Italia, Milan, Italy

Man Zhu Southeast University, Nanjing, China

Chapter 1
Enterprise Knowledge Graph: An Introduction

Jose Manuel Gomez-Perez, Jeff Z. Pan, Guido Vetere and Honghan Wu

---

```
A knowledge graph consists of a set of interconnected typed
    entities and their attributes.
```

Compared to other knowledge-oriented information systems, the distinctive features of knowledge graphs lie in their special combination of knowledge representation structures, information management processes and search algorithms. The term 'Knowledge Graph' became well known in 2012 when Google started to use knowledge graph in their search engine, allowing users to search for things, people or places, rather than just matching strings in the search queries with those in Web documents. Inspired by the success story of Google, knowledge graphs are gaining momentum in the world's leading information companies.

The idea of a knowledge graph is not completely new though. The original idea dates back to the knowledge representation technique called the Semantic Network. Later on, researchers in Knowledge Representation and Reasoning (KR) addressed

J.M. Gomez-Perez (✉)
Expert System, Prof. Waksman 10, 28036 Madrid, Spain
e-mail: jmgomez@expertsystem.com

J.Z. Pan
University of Aberdeen, King's College, Aberdeen AB24 3UE, UK
e-mail: jeff.z.pan@abdn.ac.uk

G. Vetere
IBM Italia, via Sciangai 53, 00144 Rome, Italy
e-mail: gvetere@it.ibm.com

H. Wu
King's College London, De Crespigny Park, London SE5 8AF, UK
e-mail: honghan.wu@kcl.ac.uk

J.Z. Pan et al. (eds.), *Exploiting Linked Data and Knowledge Graphs in Large Organizations*, DOI 10.1007/978-3-319-45654-6_1

some well-known issues on the Semantic Network when standardising the modern version of Semantic Network, or RDF (Resource Description Frameworks). It turns out that knowledge representation techniques, such as Knowledge Graph or Semantic Network, are useful not only for Web search, but also in many other systems and applications, including enterprise information management. The focus of the book, therefore, is about constructing, understanding and exploiting knowledge graphs in large organisations.

The basic unit of a knowledge graph is (the representation of) a singular *entity*, such as a football match you are watching, a city you will visit soon or anything you would like to describe. Each entity might have various attributes. For example, the attributes of a person include name, birthdate, nationality, etc. Furthermore, entities are connected to each other by *relations*; e.g. you *follow* one of your colleagues in Twitter. *Relations* can be used to bridge two separate knowledge graphs. For example, by saying that your Twitter ID and the ID on your driving license are denoting one and the same person, this actually interlinks Twitter data with the information space in the driver licensing agency of your country. Not surprisingly, each entity needs an identification to distinguish one another. This is the final jigsaw in the knowledge representation of knowledge graphs. Note that to facilitate the interlinking between various knowledge graphs, the entity IDs need to be *globally* unique. Types of entities and relations are defined in some machine-understandable dictionaries called ontologies. The standard ontology language is called OWL (Web Ontology Language).

The quality of a knowledge graph is crucial for its applications. For example, a knowledge graph should be consistent. In the above example, it could be the case that your contact address in your driving license is different than that in your Twitter profile. To create a knowledge graph connecting these two information spaces, such inconsistency should be resolved by keeping the correct one. In addition to consistency, one also needs to consider correctness, and coverage of knowledge graphs, as well as efficiency, fault tolerance and scalability of services based on knowledge graphs. Many of those aspects are related to, among others, the schema (ontology) of a knowledge graph.

```
A knowledge graph has an ontology as its  schema defining
    the vocabulary used in the knowledge graph.
```

1.1 A Brief History of Knowledge Graph

1.1.1 The Arrival of Semantic Networks

Knowledge management in early human history was largely shaped by oral communication before the invention of languages, which then allowed human knowledge to be recorded and passed on through generations. One of the first computer-based knowledge representation approaches are *Semantic Networks*, which represent knowledge in the form of interconnected nodes and arcs, where nodes represent objects, concepts or situations, and edges represent the relations between them, including is-a (e.g. "a chair is a type of furniture") and part-of (e.g. "a seat is part of a chair").

As regards the origin of Semantic Networks [38], some researchers argue that Semantic Networks have come from Charles S. Peirce's existential graphs, while many of them pay tribute to Quillian, who was the first to introduce Semantic Networks in his semantic memory models [194]. Semantic memory refers to general knowledge (facts, concepts and relationship), such as a chair. It is different from another kind of long-term memory, i.e. episodic memory, which relates to some specific events, such as moving a chair. After Quillian, many variants of Semantic Networks were proposed.

Compared to formal knowledge representation and reasoning formalisms, such as predicate logics, Semantic Networks are relatively easy to use and maintain. On the other hand, they suffer from some limitations. For example, there is no formal syntax and semantics for Quillian's Semantic Network. This leaves room for users to have their own interpretations of constructors in Semantic Networks, such as the is-a relation. This approach may be seen as flexible for some, but it is also criticised for making it hard to integrate Semantic Networks while preserving their original meaning. Furthermore, Semantic Networks do not allow users to define the meaning of labels on nodes and arcs.

1.1.2 From Semantic Networks to Linked Data

RDF (Resource Description Framework) is a modern standard from W3C, addressing some of the issues related to classic Semantic Networks in terms of the lack of formal syntax and semantics. For example, the is-a relation can be represented by the subClassOf property in RDF, the semantics of which is clearly defined in the RDF specifications. It should be pointed out that RDF does not address all the limitations of a Semantic Network, e.g. RDF does not allow users to define concepts either. This is, however, addressed by OWL (Web Ontology Language), a W3C standard for defining vocabularies for RDF graphs. In OWL, the part-of relation is not a built-in relation like the subClassOf property. Instead, it is a user-defined relation that can be expressed by using the existential constructor. Description Logics [18, 184] are the underpinning of the OWL standard in the Semantic Web. More details of RDF and OWL can be found in Chap. 2.

Based on RDF and OWL, Linked Data is a common framework to publish and share data across different applications and domains, where RDF provides a graph-based data model to describe objects. OWL offers a standard way of defining vocabularies for data annotations. In the Linked Data paradigm, RDF graphs can be linked together by means of mappings, including schema-level mapping (subClassOf) and object-level mapping (sameAs).

1.1.3 Knowledge Graphs: An Entity-Centric View of Linked Data

In 2012, Google popularised the term *Knowledge Graph* (KG) with a blog post titled '*Introducing the Knowledge Graph: things, not strings*',[1] while simultaneously applying the approach to their core business, fundamentally to the Web search area. Among other features, the most typical one from the user's perspective is that, in addition to a ranked list of Web pages resulting from the keyword search, Google also shows a structured knowledge card on the right, which is a small box containing a summarised information snippet about the entity that probably solves the search. Such a knowledge card contains additional information relevant to the search, contributing to relieving the burden on the user's side to pick up relevant Web pages to find answers manually. Furthermore, relations with other entities in the KG are suggested, increasing the feeling of serendipity and stimulating further exploration by the user. In most cases, such knowledge cards sufficiently fulfil searchers' information needs, significantly improving the efficiency of Web search systems both in terms of time spent per search and quality of the results.

Inspired by the successful story of Google, knowledge graphs are gaining momentum in the World Wide Web arena. In recent years, we have witnessed an increasing industrial take-up by other Internet giants, which include Facebook's Graph Search and Microsoft's Satori, continued effort made in industrial research, e.g. Knowledge Vault [69], posting community-driven events (Knowledge Graph Tutorial in WWW2015[2]; KG2014[3]), entering into academia–industry collaborations and the establishment of start-ups that specialised in areas such as Diffbot[4] and Syapse.[5] All these initiatives, taken in both academic and industrial environments, have further developed and extended the initial Knowledge Graph concept which was popularised by Google. Additional features, new insights and various applications have been introduced and, as a consequence, the notion of knowledge graphs has grown into a much broader term that encapsulates a whole line of community effort in its own right, new methods and technologies.

To explain the subtle differences between knowledge graph and Linked Data better, we first need to introduce some basic concepts. Thus, we will postpone such detailed discussions to Sect. 2.4, after providing an introduction on the background knowledge in Sects. 2.1–2.3.

[1] http://googleblog.blogspot.co.uk/2012/05/introducing-knowledge-graph-things-not.html.

[2] http://www.www2015.it/tutorials-19/.

[3] http://www.cipsc.org.cn/kg2/index_en.html.

[4] http://www.diffbot.com/products/.

[5] http://syapse.com/.

1.2 Knowledge Graph Technologies in a Nutshell

A knowledge graph based information system usually forms an ecosystem comprising three main components: construction, storage and consumption. Relevant knowledge graph technologies can be classified into one of these components of such an ecosystem where their contribution is most critical. As regards knowledge graph construction and storage, one finds technologies and tools for:

- knowledge representation and reasoning (languages, schema and standard vocabularies),
- knowledge storage (graph databases and repositories),
- knowledge engineering (methodologies, editors and design patterns),
- (automatic) knowledge learning including schema learning and population.

For the first three items, the majority of technologies are derived from the areas of KR, Databases, Ontologies and the Semantic Web. For knowledge learning, on the other hand, frameworks and technologies from Data Mining, Natural Language Processing and Machine Learning are typically employed.

From the consumption point of view, knowledge graphs' content can be directly accessed and analysed via query languages, search engines, specialised interfaces and/or generation of (domain/application-specific) graph summaries and visual analytics. In many other cases, a knowledge graph can enhance the effectiveness of a traditional information processing/access task (e.g. information extraction, search, recommendation, question answering, etc.) by providing a valuable background domain knowledge.

In this book, we cover knowledge graph technologies of all the above types, ranging from foundational representation languages like RDF to advanced frameworks for graph summarisation and question answering. Some of these technologies are useful for understanding knowledge graphs, while others help in exploiting knowledge graphs to support intelligent systems and applications.

1.3 Applications of Knowledge Graphs for Enterprise

Back in 2008, ongoing and future trends in semantic technologies were forecast to lie at the intersection of three main dimensions:

- natural interaction,
- the Web 2.0,
- service-oriented architectures.

If we abstract away from those particular terms, the actual meaning becomes quite simple:

- ease of *access* to computer systems by end users,

- *empowerment* of user communities to represent, manage and share knowledge in collaborative ways,
- machine *interoperability*.

Since then, countless research challenges have been faced in areas such as Knowledge Acquisition, Representation and Discovery, Knowledge Engineering Methodologies, Vocabularies, Scalable Data Management Architectures, Human–Computer Interaction, Information Retrieval and Artificial Intelligence, where semantic technologies have been involved, contributing to crucial advances in knowledge-intensive systems.

Now, like then, the value of data as the driving force behind intelligent applications remains. However, there is a new trend gaining momentum, which lies at the realisation that such a *value is directly proportional to the interlinkedness of the data* not only in complex, open-ended systems like the Web but also in specific enterprise applications based on combinations of both corporate and open data. More suited to look-up and relatedness operations, poorly formalised but highly interconnected data are becoming more popular than highly formalised but isolated datasets. The current application landscape, more oriented towards mobile and real time, is enforcing this new paradigm shift.

Google understood this very well and in 2012 started driving this trend in the industry by releasing their Knowledge Graph as a way to master such value, a large knowledge base that enhances its search engine's results with semantic-search information gathered from a wide variety of sources. Interestingly, the Knowledge Graph provides a way to connect the dots (entities) by means of explicit relations, with both entities and relations described following formal (but lightweight) models and reusing existing datasets like Freebase. After Google, other knowledge graphs arrived at the Internet scale, including those of Microsoft and Yahoo! Nowadays, it is the turn of enterprises and public administrations to leverage the Knowledge Graph concept at a corporate level in order to describe their data, enrich it by interlinking it with other knowledge bases both within and outside their environment and revitalise the development of knowledge-intensive systems on top of it.

Compared to 2008 [24], the interest in Market Intelligence and data-intensive sectors[6] and the role of knowledge graphs have increased dramatically while others, like corporate knowledge management and open government, are still there, though with slightly different foci. Next, we give an account of some selected applications that use knowledge graphs in such sectors, which will hopefully provide insight into the potential impact and future opportunities of knowledge graphs.

Corporate Knowledge Management

Open Innovation

Nowadays, especially after the recent financial downturn, companies are looking for much more efficient and creative business processes so as to place better solutions in the market in less time with less cost. There is a general impression that communication and collaboration, especially mixed with Web 2.0 approaches within companies

[6]IDG Enterprise Big Data report—http://www.idgenterprise.com/report/big-data-2.

and ecosystems (so-called Enterprise 2.0 [156]), can boost the innovation process with positive impacts on business indicators.

Open innovation [45] within an Enterprise 2.0 context is one of the most popular paradigms for improving the innovation processes of enterprises, based on the collaborative creation and development of ideas and products. The key feature of this new paradigm is that knowledge is exploited in a collaborative way flowing not only between internal sources, e.g. R&D departments, but also between external ones such as employees, customers, partners, etc. In this scenario, corporate knowledge graphs can be used to (i) support the semantic contextualisation of content-related tasks involving individuals and roles and (ii) help in discovering relations between communities of employees, customers and providers, with shared knowledge and interests.

The introduction of the open innovation paradigm in an enterprise entails not just a modification of corporate innovation processes but also a cultural change which requires support by an advanced technological infrastructure. Corporate knowledge has to be made explicit, exchanged and shared between participants, and therefore tools for knowledge management, analysis support and information structuring are required to make these tasks affordable and the knowledge available to all the involved actors. In addition, tools supporting the innovation process need to provide a high degree of interactivity, connectivity and sharing. In a scenario where collaborative work is not supported and members of a community could barely interact with each other, solutions to everyday problems and organisational issues rely on an individual's initiative. Innovation and R&D management are complex processes for which collaboration and communication are fundamental. They imply creation, recognition and articulation of opportunities, which need to be evolved into a business proposition at a subsequent stage. Interactivity, connectivity and sharing are the features to consider when designing a technological framework for supporting collaborative innovations [90]. All these characteristics can be identified in Enterprise 2.0 environments.

However, Enterprise 2.0 tools do not provide formal models which are used to create complex systems that manage large amounts of information. This drawback can be overcome by incorporating corporate knowledge graphs introducing computer-readable, interlinked representations of entities. Open innovation platforms similar to the one described in [1] leverage the concept of a corporate knowledge graph to relate people, interests and ideas in a corporate knowledge management environment throughout sectors, involving employees, clients and other stakeholders.

The impact of knowledge graphs through their application in open innovation is illustrated by their adoption in large corporations belonging to several sectors such as banking, energy and telecommunications (see further details in [45]), with companies such as Bankinter, Repsol and Telefonica, which have positioned themselves at the forefront of these efforts. What all these efforts have in common is the need to connect innovative ideas and people in order to orchestrate a healthy innovation ecosystem, addressing several challenges, like:

- Handling the information created by thousands of employees,
- evaluating their ideas efficiently,
- reducing false positives (ideas that reach the market and fail) and false negatives (valuable ideas which are rejected even before they can reach the market),
- stimulating the communication among people located around the globe, in different languages.

Intra-enterprise Micro-knowledge Management

As seen above, knowledge management is one of the key strategies that allow companies to fully tap into their collective knowledge. However, two main entry barriers usually limit the potential of this approach: (i) the barriers that employees encounter discouraging them from strong and active participation (knowledge providing) and (ii) the lack of truly evolved intelligent technologies that enable employees to easily benefit from the global knowledge provided by the companies and other users (knowledge consuming). In [188], miKrow, a lightweight framework for knowledge management, was proposed based on the combination of two layers that exploit corporate knowledge graphs to cater to both needs: a microblogging layer that simplifies how users interact with the whole system and a semantic engine that performs all the intelligent heavy lifting by combining semantic indexing and search of microblogs and users.

The miKrow interaction platform is a Web application that is designed as per the Web 2.0 principles of participation and usability. miKrow centres interaction around a simple text box user interface with a single input option for end users, where they are able to express what they are doing, or more typically in a work environment, what they are working at. This approach diverges from classical KM solutions which are powerful yet complex, following the idea of simplicity behind the microblogging paradigm in order to reduce the general entry barriers for end users. The message is semantically indexed against the underlying knowledge graph so that it can be retrieved later, as well as the particular worker linked to it. miKrow's semantic functionalities are built on top of the underlying knowledge graph, which captures and relates the relevant corporate entities.

Market Intelligence

According to the consulting company International Data Corporation (IDC) in its 2014 IDG Enterprise Big Data report, on an average enterprises spent $8M on leveraging value out of data in 2014, with penetration levels of 70 % and 56 % for large enterprises and SMEs, respectively. Improving the quality of the decision-making process (59 %), increasing the speed of decision-making processes (53 %), improving planning and forecasting (47 %), and developing new products/services and revenue streams (47 %) are the top four areas accelerating investment in data-driven business initiatives.

This trend is especially acute in the digital content and advertising sector. The communication between brands and consumers is set to explode. Product features are no longer the key to sales and the combination of both personal and collective benefits is becoming an increasingly crucial aspect. As a matter of fact, brands providing such value achieve a higher impact and consequently derive clearer economic benefits. On the other hand, millennials [98] are taking over, inducing a dramatic change in the way consumers and brands engage and what channels and technologies are required to enable the process. As a result, traditional boundaries within the media industry are being stretched and new ideas, inventions and technologies are needed to keep up with the challenges raised by the increasing demands of this data-intensive, in-time, personalised and thriving market.

HAVAS, the fourth largest media group worldwide, seeks to interconnect start-ups, innovators, technology trends, other companies and universities worldwide in one of the first applications of Web-scale knowledge graph principles to the enterprise world and media [46]. The resulting enterprise knowledge graph supports analytics and strategic decision-making for the incorporation of such talent within their first 18 months life span. Such an endeavour involves the application of semantic technologies by extracting start-up information from online sources, structuring and enriching it into an actionable, self-sustainable knowledge graph, and providing media businesses with strategic knowledge about the most trending innovations. While the previous success stories deal with the management of corporate knowledge within corporations, in this case the focus lies in creating competitive intelligence.

As we already know, innovation is often misunderstood and difficult to integrate into corporate mind-set and culture. So, why not activate relevant external talent and resources when necessary? The discovery and surveillance of trends and talent in the start-up ecosystem can be time consuming, though. HAVAS' knowledge graph sets its semantic engineering to run a surveillance monitoring of the entrepreneurial digital footprint, collecting and gathering fruitful insight and information, which provides the staff with clear leads for analysis. By automating part of the research process, analysts can get there faster and more accurately than competitors, leveraging millions of data points, and implementing consistency through a single and shared knowledge entry point. At the moment the knowledge graph is being opened to HAVAS' network, with teams in 120 offices around the world and clients, providing access to knowledge about the best-in-class talent to implement new thinking and cutting-edge solutions to the never-ending and evolving challenges within the media industry. Based on the knowledge graph, teams also rate and share experiences, ensuring that learning can be propagated across the network.

IBM Watson

IBM Watson is a cognitive computing platform available in the cloud, developed by IBM as an outcome of the *Jeopardy!* Q&A challenge[7]; cf. Sect. 7.2 and the Foreword of this book by Chris Welty. Watson uses Natural Language Processing and Machine Learning to discover insights from large amounts of unstructured data and provides a variety of services to work with this knowledge. Knowledge Graphs (such as Prismatic, DBPedia and YAGO) were at the core of the IBM's Q&A system.[8] IBM Watson services available today provide KGs capabilities through many services

[7] http://www.ibm.com/smarterplanet/us/en/ibmwatson/.

[8] IBM Journal of Research and Development, Vol. 56, No. 3/4, May/July 2012.

and application program interfaces (APIs), such as the Watson Concept Expansion and Insight.[9] Ongoing research and development aim at extending the availability of large structured knowledge bases to Dialog Services and other cognitive front ends.

1.4 How to Read This Book

1.4.1 Structure of This Book

This book introduces the key technologies for constructing, understanding and exploiting knowledge graphs. We hope you like reading this chapter so far. The rest of this book contains three parts, as illustrated in Fig. 1.1 (p. 11):

- **Part 1** contains Chaps. 2 and 3, in which we first introduce some basic background knowledge and technologies, and then present a simple architecture in order to help you to understand the main phases and tasks required during the lifecycle of knowledge graphs.
  - **Chapter** 2 introduces the background knowledge for studying and understanding the Knowledge Graph. Furthermore, we include a bit more discussion in the end to clarify the relations between Knowledge Graphs and Linked Data, as well as different purposes of building knowledge groups, e.g. for Web search versus for enterprise information systems.
  - **Chapter** 3 introduces a three-layer architecture of the Knowledge Graph application: (L1) Acquisition and Integration Layer; (L2) Knowledge Storing and Accessing Layer; and (L3) Knowledge Consumption Layer.
- **Part 2** is the main technical part for the Knowledge Graph, which contains Chaps. 4–7.
  - **Chapters** 4 **and** 5 further explain the layer L1 and address the state-of-the-art technology of knowledge acquisition and ontology construction.
  - **Chapters** 6 **and** 7 further explain the layer L3, where Chap. 6 introduces the key technologies of summarisation service, while Chap. 7 introduces the techniques of applying knowledge graphs in question answering (like the IBM Watson DeepQA).

  Based on the level of technical details, we have placed an asterisk on the titles of some chapters and sections, which contain detailed technical descriptions (e.g. formal definitions or formulas) or advanced topics (e.g. statistical/logical reasoning). Specifically, they are Chap. 5, Sects. 6.4 and 7.4.

[9]http://www.ibm.com/smarterplanet/us/en/ibmwatson/developercloud/.

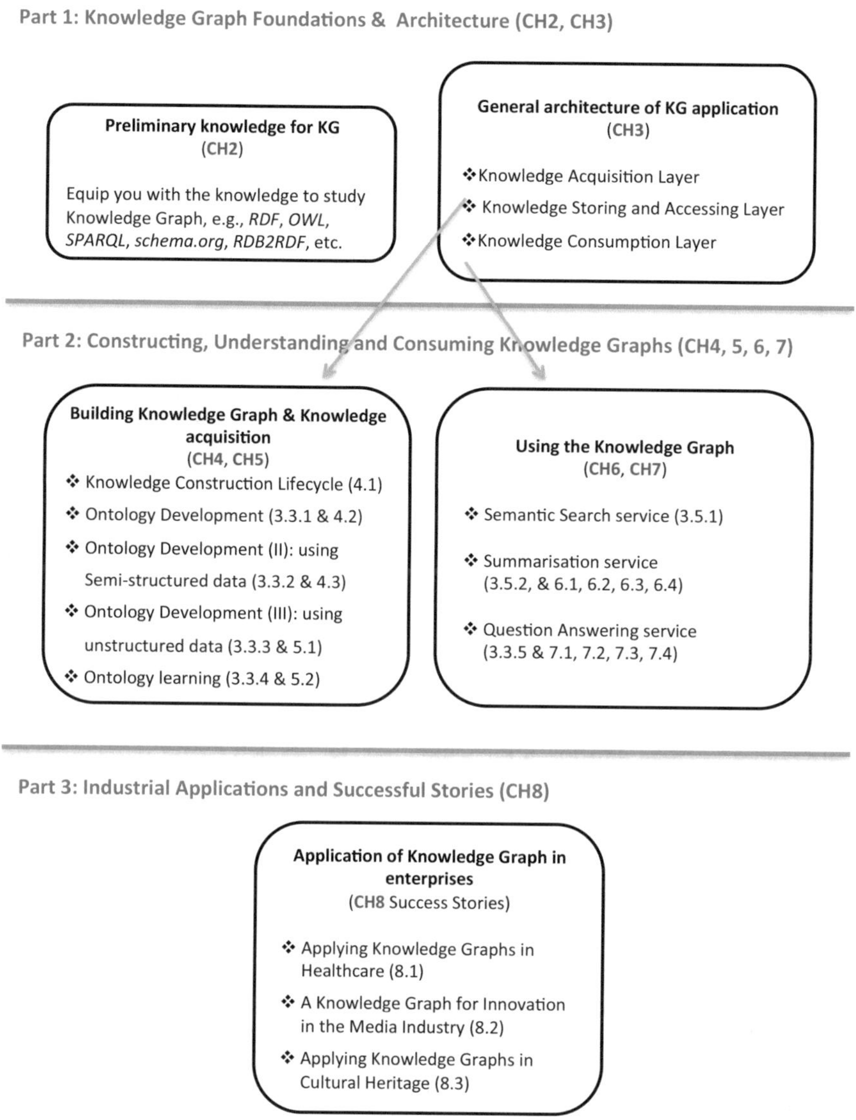

Part 1: Knowledge Graph Foundations & Architecture (CH2, CH3)

Preliminary knowledge for KG
(CH2)

Equip you with the knowledge to study Knowledge Graph, e.g., *RDF*, *OWL*, *SPARQL*, *schema.org*, *RDB2RDF*, etc.

General architecture of KG application
(CH3)

- Knowledge Acquisition Layer
- Knowledge Storing and Accessing Layer
- Knowledge Consumption Layer

Part 2: Constructing, Understanding and Consuming Knowledge Graphs (CH4, 5, 6, 7)

Building Knowledge Graph & Knowledge acquisition
(CH4, CH5)

- Knowledge Construction Lifecycle (4.1)
- Ontology Development (3.3.1 & 4.2)
- Ontology Development (II): using Semi-structured data (3.3.2 & 4.3)
- Ontology Development (III): using unstructured data (3.3.3 & 5.1)
- Ontology learning (3.3.4 & 5.2)

Using the Knowledge Graph
(CH6, CH7)

- Semantic Search service (3.5.1)
- Summarisation service (3.5.2, & 6.1, 6.2, 6.3, 6.4)
- Question Answering service (3.3.5 & 7.1, 7.2, 7.3, 7.4)

Part 3: Industrial Applications and Successful Stories (CH8)

Application of Knowledge Graph in enterprises
(CH8 Success Stories)

- Applying Knowledge Graphs in Healthcare (8.1)
- A Knowledge Graph for Innovation in the Media Industry (8.2)
- Applying Knowledge Graphs in Cultural Heritage (8.3)

Fig. 1.1 The three parts of the main content of this book

- **Part 3** (Chap. 8) introduces the successful stories of applying Knowledge Graph in Healthcare (8.1), Media Industry (8.2) and Cultural Heritage (8.3).

In Chap. 9 we conclude this book which shares some valuable experience of the editors and authors about their works on knowledge graphs.

1.4.2 Who This Book Is For

This book is for academic researchers, knowledge engineers and IT professionals who are interested in acquiring industrial experience in using knowledge graphs for enterprises and large organisations. The book provides readers with an updated view of methods and technologies related to knowledge graphs, including illustrative corporate use cases.

I am an academic researcher/postgraduate student, what can I learn from this book?

For readers who are familiar with semantic technologies, this book provides an overview of the state of the art in knowledge graph technologies and of research methods and tools involved in building, managing and exploiting knowledge graphs. Readers will also benefit from insight and lessons learnt from the application of such approaches to different real-life problems and corporate environments.

I am an engineer or a manager in industry, what can I learn from this book?

Readers from industry will find in this book an open door to new and effective means to structure knowledge and link the different corporate assets in a way that modern organisations can exploit efficiently for a number of different purposes, including knowledge management and decision-making. Knowledge and software engineers will become familiar with the relevant techniques in the area while managers will find additional insight into how this paradigm can unlock new business opportunities to exploit both corporate and publicly available knowledge.

1.4.3 How to Use This Book

The content of this book is structured in three parts: the preliminary fundamental knowledge, the key technologies of Knowledge Graph and the applications.

Figure 1.1 (p. 11) can be used as a road-map across this book to remind the readers where they are in the journey, and help them to skip some sections, for example some sections are too technical to be of interest to general readers, and to find the most important content for them.

In the following, we provide a few details of each chapter.

Chapter 1 (Enterprise Knowledge Graph: An Introduction)

briefly explains why the editors and the authors presented this book. As mentioned in the title, this book is about how knowledge graphs are used in enterprises as knowledge management methods. In this chapter, it firstly introduces the brief evolutional history of Knowledge Graph and the key technologies used in it. Then it introduces the main applications of Knowledge Graph in enterprises. A guidance of how to read this book is also provided.

Chapter 2 (Knowledge Graph Foundations)

presents a high-level overview of the foundations of knowledge graphs. We want to introduce, in a very light way, all the concepts and basics we need for understanding and working with knowledge graphs. We start by describing how knowledge is represented and the query languages that are under the hood of the knowledge graphs. Next, we briefly present the models/vocabularies/ontologies that are needed for describing knowledge. Finally, we introduce a few basic transformation approaches from the original data source formats.

Chapter 3 (Knowledge Architecture for Organisations)

introduces a high-level overview of what is needed in order to create, maintain and exploit knowledge graphs. We realise that there is no one way of doing this for all organisations and all use cases of knowledge graphs; hence in this chapter, we introduce an abstract reference architecture that includes the main phases and tasks required during the lifecycle of knowledge graphs. For each of the phases and tasks, we then present a more detailed description of possible approaches, methodologies and tools which have been reported in the literature. By the end of this chapter you should have a good idea of the tasks you would likely encounter when building and maintaining knowledge graphs. You would also have a better understanding of how knowledge graphs can be used within large organisations.

Chapter 4 (Construction of Enterprise Knowledge Graphs (I))

as well as Chap. 5, focuses on the *Acquisition and Integration Layer* of Chap. 3's reference architecture. In particular, we start with a generic lifecycle for constructing and maintaining knowledge graphs in Sect. 4.1 and, then, we elaborate on the knowledge graph construction approaches a.k.a. the modelling and data lifting steps in the lifecycle. In this chapter, we focus on supervised approaches to constructing new knowledge, i.e. approaches involving human effort. Specifically, for the modelling step we introduce a competency question based ontology authoring framework, while for data lifting we discuss a semi-automated approach for creating linkages among heterogeneous data sources.

Chapter 5 (Construction of Enterprise Knowledge Graphs (II))

continues with the *Acquisition and Integration Layer* of Chap. 3 on a reference architecture, focusing on knowledge graph construction techniques. Nevertheless, we shift from semi-automated approaches to automated approaches of knowledge graph construction by describing two additional frameworks, one for entity/scope resolution of textual data (Sect. 5.1) and one for the learning of ontological schemas from data (Sect. 5.2).

Chapter 6 (Understanding Knowledge Graphs)

identifies and introduces a set of techniques that make knowledge graphs directly available to end users. Among others, we lay a special focus on knowledge graph understanding techniques, many of which were especially designed for scenarios in large organisations.

Chapter 7 (Question Answering and Knowledge Graphs)

This is a "star" chapter. We primarily consider the tasks of question answering over text documents (Sect. 7.1) and knowledge graphs (Sect. 7.2), and we present an overview of relevant methodologies, technologies and systems. Moreover, in Sect. 7.3, we describe a state-of-the-art question-answering system that combines knowledge coming from the text analysis and knowledge graphs.

Chapter 8 (Success Stories)

presents success stories of the applications of Knowledge Graph techniques from various domains (healthcare, media and culture) and different organisations (international company—IBM, Small and Medium Enterprises—HAVAS, and University—the University of Aberdeen).

Chapter 9 (Conclusion and Outlook)

concludes the book with a brief review of the whole book. Furthermore, it shares the valuable experience from the editors and authors of "things to keep in mind" when adopting knowledge graphs.

We hope you will enjoy "Exploiting Linked Data and Knowledge Graphs in Large Organizations", as we have been enjoying it.

Part I
Knowledge Graph Foundations & Architecture

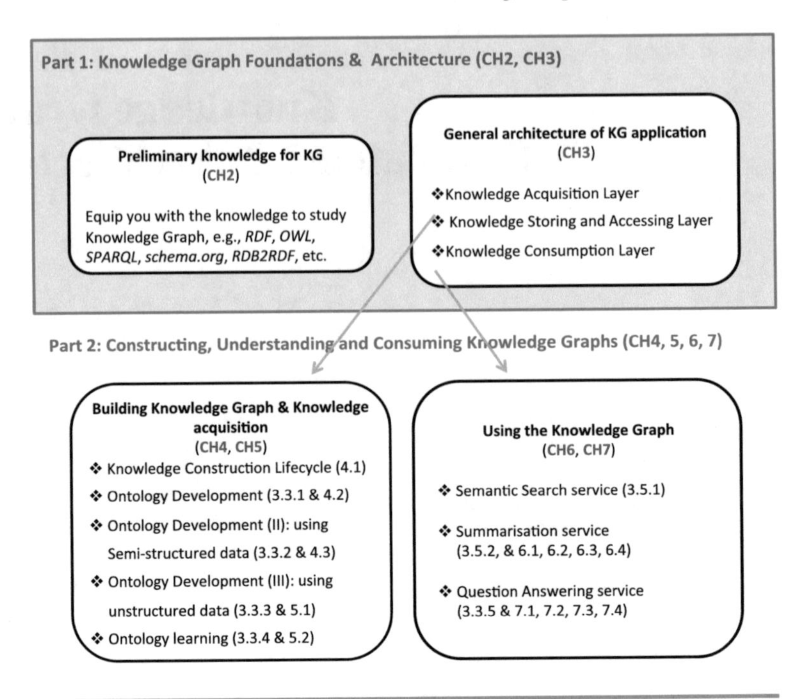

Part 3: Industrial Applications and Successful Stories (CH8)

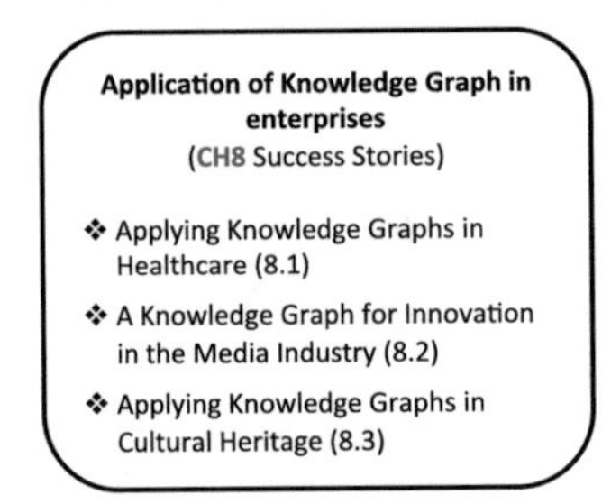

Fig. RoadMap. 1 The roadmap of Part I

In the first part of this book, first we introduce some basic background knowledge and technologies about building and using knowledge graphs. Then we present an ARA architecture in order to help you to understand the main phases and tasks required during the lifecycle of knowledge graphs.

For readers who are familiar with semantic technologies, e.g. *RDF*, *OWL*, *SPARQL*, *schema.org*, *RDB2RDF*, could jump directly to Section 2.4.

Part I contains the following chapters:
Chapter 2: Knowledge Graph Foundations
Chapter 3: Knowledge Architecture for Organisations

Chapter 2
Knowledge Graph Foundations

Boris Villazon-Terrazas, Nuria Garcia-Santa, Yuan Ren, Alessandro Faraotti, Honghan Wu, Yuting Zhao, Guido Vetere and Jeff Z. Pan

This chapter presents a high-level overview of the foundations of knowledge graphs. The goal of this chapter is to introduce, in a light way, concepts and basics that we need for understanding and working with knowledge graphs. This chapter starts by introducing some standards on representing knowledge graphs, including their schemas called ontologies, as well as some widely used vocabularies defined in ontologies. This chapter also introduces standards on querying knowledge graphs as well as transforming data in other formats into knowledge graphs. As the book is about knowledge graph for enterprise rather than for Web search, we conclude this chapter by comparing these two scenarios, so as to set the scene for the rest of the book.

B. Villazon-Terrazas (✉) · N. Garcia-Santa
Expert System, Prof. Waksman 10, 28036 Madrid, Spain
e-mail: bvillazon@expertsystem.com

N. Garcia-Santa
e-mail: ngarcia@expertsystem.com

A. Faraotti · G. Vetere
IBM Italia, via Sciangai 53, 00144 Rome, Italy
e-mail: alessandro.faraotti@it.ibm.com

G. Vetere
e-mail: gvetere@it.ibm.com

H. Wu
King's College London, De Crespigny Park, London SE5 8AF, UK
e-mail: honghan.wu@kcl.ac.uk

Y. Zhao
IBM Italia, Circonvallazione Idroscalo, 20090 Milan, Italy
e-mail: yuting.zhao@it.ibm.com

Y. Ren · J.Z. Pan
University of Aberdeen, King's College, Aberdeen AB24 3UE, UK
e-mail: y.ren@abdn.ac.uk

J.Z. Pan
e-mail: jeff.z.pan@abdn.ac.uk

J.Z. Pan et al. (eds.), *Exploiting Linked Data and Knowledge Graphs in Large Organizations*, DOI 10.1007/978-3-319-45654-6_2

2.1 Knowledge Representation and Query Languages

Knowledge Representation and Reasoning (KR) is the field of Artificial Intelligence (AI) dedicated to representing information of the world in the form so as to allow a computer to reason automatically with relevant information for solving complex tasks such as decision support. As discussed in Chap. 1, knowledge graph can be seen as some modern variant of KR formalism called the Semantic Network.

A knowledge graph is a set of typed entities (with attributes) which relate to one another by typed relationships. The types of entities and relationships are defined in schemas that are called ontologies. Such defined types are called vocabulary. In this section, we will introduce the standard (RDF) for representing knowledge graphs, two standards for defining ontologies (RDFS and OWL) and the standard for querying knowledge graphs. Note that following standards does not prevent one from using some customised serialisations for building enterprise knowledge graphs.

2.1.1 RDF and RDFS

In this section, we introduce RDF (Resource Description Framework) and RDFS (RDF Schema). As discussed in Sect. 1.1, RDF is the modern standard for Semantic Networks. RDFS is a simple schema language for RDF knowledge graphs. In Sect. 2.1.2, we will introduce another standard OWL that offers a more comprehensive family of schema languages for knowledge graphs.

RDF

RDF is a Recommendation (standard) from the World Wide Web Consortium (W3C), for describing entities, often referred to as resources in W3C. A resource can be anything we can identify, such as a person, a homepage or great dragons in the Game of Thrones. In this section, we will use some examples of organisations taken from the New York Times Linked Dataset[1] and DBpedia[2] for illustration.

RDF Triples and Graphs

Resources are described in RDF *triples*, also known as *statements*, with predicate-value pair as follows. Values in RDF triples can be resources as well (Fig. 2.1).

---

[1]http://datahub.io/dataset/nytimes-linked-open-data.

[2]http://dbpedia.org/About.

```
[subject  predicate object .]
```

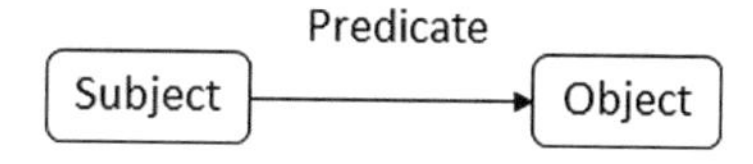

Fig. 2.1 An RDF statement

An RDF graph consists of a set of triples. It can be visualised as a node-directed arc diagram, in which each triple is represented as a node-arc-node link. There can be three kinds of nodes in an RDF graph: Internationalised Resource Identifiers, or IRIs, literals and blank nodes [61]. More precisely, in an RDF triple,

- the subject can be an IRI or a blank node;
- the predicate should be an IRI;
- the object can be an IRI, a literal or a blank node.

An IRI (Internationalised Resource Identifier) within an RDF graph is a string that univocally identifies a resource [71]. IRIs are a generalisation of URIs (Uniform Resource Identifiers) [27] that permit a wider range of characters.

Literals are used for values such as strings, numbers and dates. A literal in an RDF graph consists of two or three elements:

- a *lexical form*, which is a Unicode string and should be in Normal Form C [65];
- a *datatype IRI*, which is an IRI identifying a datatype that determines how the lexical form parts to a literal value, and
- if the datatype IRI is http://www.w3.org/1999/02/22-rdf-syntax-ns#langString, a nonempty *language* tag as defined by [189].

Blank nodes are disjoint from IRIs and literals. Other than that, the set of possible blank nodes is arbitrary. RDF makes no reference to any internal structure of blank nodes. A blank node is a node in an RDF graph representing a resource for which an IRI or literal is not given. The resource represented by a blank node is also called an anonymous resource. Blank nodes are treated as simply indicating the existence of a thing, without using an IRI to identify any particular thing. Blank nodes give capability to:

- describe multicomponent structures, like the RDF containers;
- describe reification, i.e. provenance information;
- represent complex attributes without having to explicitly name the auxiliary node, e.g. the address of a person consisting of the street, number, postal code and city;
- offer protection of the inner information, e.g. protecting the sensitive information of the customers from browsers.

Figure 2.2 presents an RDF graph with one triple. The subject is http://dbpedia.org/resource/Bolivia, the predicate is http://dbpedia.org/ontology/longName and the

Fig. 2.2 An example of a triple

object is the literal *"Plurinational State of Bolivia"*. This triple says that Bolivia's long name is Plurinational State of Bolivia.

IRIs often begin with a common substring known as a *namespace IRI*. Some namespace IRIs are associated by convention to a short name known as a *namespace prefix*. Prefixes can be declared as follows.

```
prefix dbpedia: http://dbpedia.org/resource/
prefix dbpedia-owl: http://dbpedia.org/ontology/
prefix xsd: http://www.w3.org/2001/XMLSchema#
```

Figure 2.3 depicts the previous RDF graph (Fig. 2.2) using the declared prefixes. Moreover, we can revise the graph and include a type for the literal, see Fig. 2.4. Finally, we can also include a language description to our literal, see Fig. 2.5.

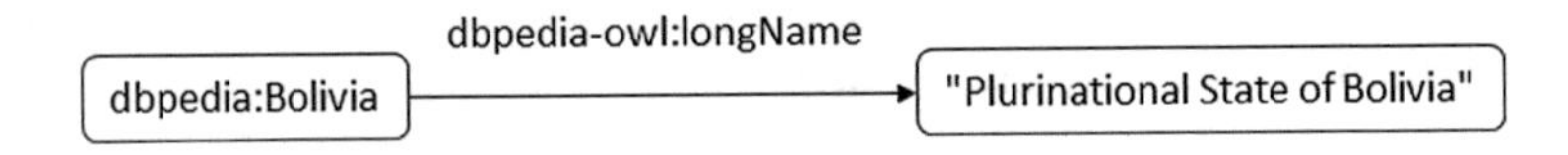

Fig. 2.3 An example of a triple using prefixes

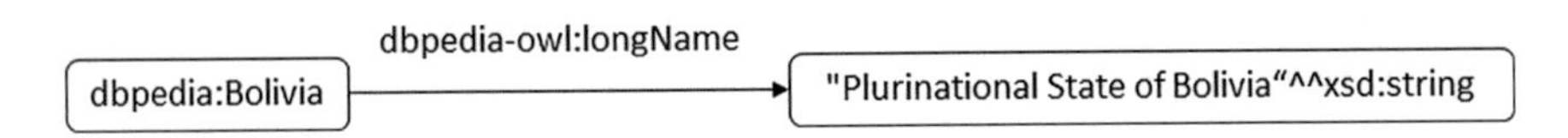

Fig. 2.4 An example of a triple using typed literals

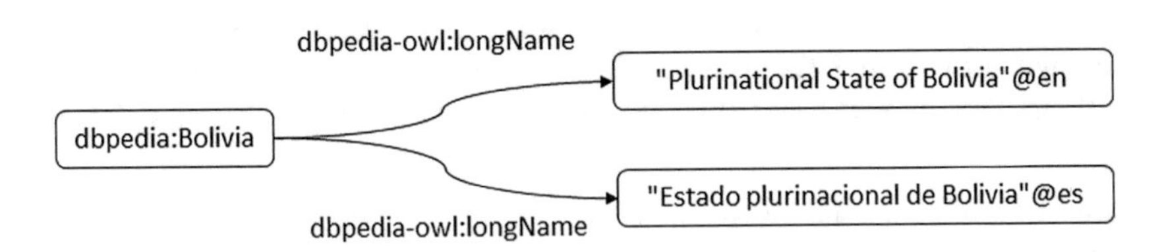

Fig. 2.5 An example of a triple using language tags

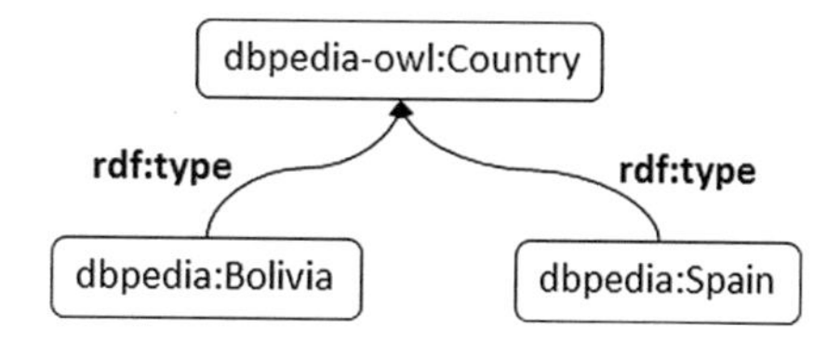

Fig. 2.6 An example of rdf:type for connecting instances and classes

Unlike classic Semantic Networks (cf. Sect. 1.1), RDF has some language level predefined properties. Here we will introduce the rdf:type, while the other type will be explained later on in this section when we introduce the RDF Schema. The `rdf:type` property is used to classify resources in categories/classes. It is the is-a relationship in Semantic Networks. Figure 2.6 depicts an example of the use of rdf:type.

Serialising RDF

RDF provides the standard data model for a knowledge graph. There are several serialisation syntaxes for storing and exchanging RDF, such as Turtle [190], RDF/XML [3], RDFa [4], N-Triples [2], NQUADS [1] and JSON-LD [220].

A Turtle document allows writing down an RDF in a compact textual form. Comments may be given after # that is not part of another lexical token. IRIs should be enclosed in <>.

```
@prefix foaf: <http://xmlns.com/foaf/0.1/> #this is a
    declaration of  the prefix (foaf) for the Friend of A
    Friend vocabulary.
```

Here are some examples of literals, including a long literal with more than one line.

```
"Literal"
"Literal"@language
"""Long literal with
multiple lines"""
```

And these are the examples of datatyped literals, having the "lexical form"^^ datatypeURI.

```
"10"^^xsd:integer
"2006-09-04"^^xsd:date
```

Next, we provide some examples for triples and abbreviations. Triples are ended with a "." like the following two.

```
lab:aleix foaf:knows lab:nuria .
lab:aleix foaf:knows lab:almu .
```

One could use "," to separate different values for triples with the same subject and predicate, like the above two, as follows.

```
lab:aleix foaf:knows lab:nuria , lab:almu
```

Similarly, one could use ";" to separate different predicate-values pairs for triples with the same subject, like the two below

```
lab:aleix foaf:nickname "paco" .
lab:aleix foaf:currentProject lab:K-Drive .
```

as follows:

```
lab:aleix foaf:nickname "paco" ; foaf:currentProject lab:K-
    Drive .
```

Now with all the basics introduced, let us have a look at some of the RDF graphs mentioned earlier. For example, the RDF graph depicted in Fig. 2.3 can be serialised in Turtle as follows:

```
@prefix dbpedia: <http://dbpedia.org/resource/>
@prefix dbpedia-owl: <http://dbpedia.org/ontology/>

dbpedia:Bolivia dbpedia-owl:longName "Plurinational State
    of Bolivia" .
```

The RDF graph depicted in Fig. 2.5 can be serialised in Turtle as follows:

```
@prefix dbpedia: <http://dbpedia.org/resource/>
@prefix dbpedia-owl: <http://dbpedia.org/ontology/>

dbpedia:Bolivia dbpedia-owl:longName "Plurinational State
    of Bolivia"@en  .
dbpedia:Bolivia dbpedia-owl:longName "Estado plurinacional
    de Bolivia"@es .
```

RDFS

RDFS provides a simple schema language for RDF, and allows one to declare classes/properties, using the predefined language level class *rdfs:Class*/property *rdfs:Property*; see the following example of the declaring org:Organization as a class and org:hasHomePage as a property:

```
org:Organization  rdf:type rdfs:Class .
org:Start-up a rdfs:Class . #we can also replace ''rdf:type
    '' with  its abbreviation ''a''
org:hasHomePage  rdf:type rdfs:Property .
```

In addition, RDFS can also specify some dependencies among classes and properties, using the predefined language level properties *rdfs:subClassOf, rdfs:subPropertyOf, rdfs:domain* and *rdfs:range*:

- [C1 rdfs:subClassOf C2 .] This says C1 is a subclass of C2, meaning all instances of C1 will also be instances of C2. The following example says that org:Start-up is a subclass of org:Non-GovOrganization.

```
org:Start-up rdfs:subClassOf org:Organization .
```

- [P1 rdfs:subPropertyOf P2 .] This says P1 is a sub-property of P2, meaning all instances of P1 will also be instances of P2. The following example says that org:hasEnglishHomePage is a sub-property of org:hasHomePage.

```
org:hasEnglishHomePage rdfs:subPropertyOf  org:
    hasHomePage .
```

- [P domain C.] This says the property P has domain C, meaning every resource that can have property P is an instance of the class C. The following example says that every resource that can have org:hasPersonalHomePage is a sub-property of org:hasHomePage.

```
org:hasHomePage rdfs:domain  org:Organization .
```

- [P range E.] This says the property P has range C, meaning every value of the property P is an instance of the E, where E can be either a class or a datatype (such as rdfs:Literal). The following example says that every value org:hasPersonalHomePage is an instance of rdfs:Literal.

```
org:hasHomePage rdfs:range   rdfs:Literal .
```

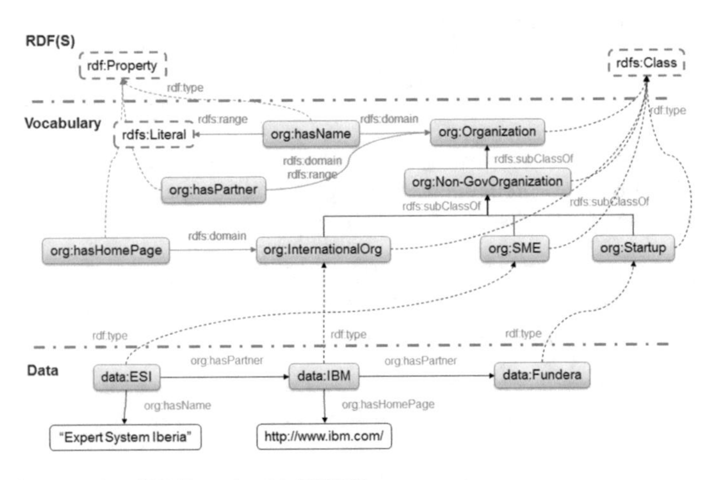

Fig. 2.7 An example of RDF graph with RDF(S) components

The RDF graph depicted in Fig. 2.7 can be serialised in Turtle as follows.

```
@prefix org: <http://www.w3.org/ns/org#>
@prefix rdf: <http://www.w3.org/1999/02/22-rdf-syntax-ns#>
@prefix rdfs: <http://www.w3.org/2000/01/rdf-schema#>
@prefix data: <http://data.lab.expertsystem.com/>

org:Organization a rdfs:Class .

org:Non-GovOrganization a rdfs:Class ;
            rdfs:subClassOf org:Organization .

org:InternationalOrg a rdfs:Class ;
          rdfs:subClassOf org:Non-GovOrganization .

org:SME a rdfs:Class ;
    rdfs:subClassOf org:Non-GovOrganization .

org:Startup a rdfs:Class ;
      rdfs:subClassOf org:Non-GovOrganization .

org:hasName a rdf:Property ;
      rdfs:domain org:Organization;
      rdfs:range rdfs:Literal.

org:hasPartner a rdf:Property ;
      rdfs:domain org:Organization ;
      rdfs:range org:Organization .

org:hasHomePage a rdf:Property ;
      rdfs:domain org:InternationalOrg .

data:ESI a org:SME ;
      org:hasPartner data:IBM ;
      org:hasName "Expert System Iberia" .

data:IBM a org:InternationalOrg ;
      org:hasPartner data:Fundera ;
      org:hasHomePage "http://www.ibm.com" .

data:Fundera a org:Start-up .
```

It should be noted that RDFS only provides limited expressive power as a schema language. Here, we briefly discuss its limitations:

1. It does not support negation; e.g. we could say data:ESI is an org:SME but cannot express that data:ESI is not a person in RDF.
2. It does not provide constructors to define classes; e.g. we cannot define what org:SME is in RDFS.
3. Although it provides schema-level alignments (with rdfs:subClassOf and rdfs:subPropertyOf), it does not support instance-level alignment; e.g. it cannot express that data:ESI is the same as db:ESI.

All the above limitations are addressed by the more comprehensive schema language OWL, which will be introduced in the next section.

2.1.2 OWL

While RDF is the modern standard for Knowledge Graph/Semantic Network, the standard ontology language OWL is based on a family of formal knowledge representations called *Description Logics* (DLs) [16]. The first DL is called KL-One, which provides more expressive power than the Semantic Network while remaining decidable. The latest version of OWL is OWL 2 [187], which has been recommended by the W3C as the de facto standard for Web ontologies. Reasoning services of OWL ontologies can be used to check the logical and semantic inconsistencies of a knowledge graph, as well as support query answering and question answering that we will discuss later on in the book.

Syntactically OWL can be regarded as an extension of RDFS with additional vocabulary predefined by the OWL schema.[3] This schema vocabulary provides extensively high expressive power for people to construct ontologies and/or to annotate their data, such as qualified cardinality restriction, property chain, self-restriction, symmetric and/or reflexive property. In what follows, we show some examples from the Travel ontology[4] to illustrate the expressive power offered by OWL 2.

Example 1 Below is a snippet of the OWL 2 file about a property:

```
:borders rdf:type owl:ObjectProperty ,
                  owl:SymmetricProperty ;
         rdfs:domain :AdministrativeDivision ;
         rdfs:range :AdministrativeDivision ;
         owl:propertyChainAxiom ( :hasBoundary
                                  :boundaryOf
                                ) .
```

It says that in travel ontology, an *object property* called *boarders* is specified between two instances of *AdministrativeDivision*. There are a few interesting features of this property which are not available in RDF(S):

1. The property has type `owl:SymmetricProperty`, indicating that if an entity `a` boarders another entity `b`, then `b` also boarders `a`.
2. The property is asserted to have an `ow:propertyChainAxiom` with another two properties `hasBoundary` and `boundaryOf`. This implies that, if `a` has boundary `b`, and `b` is a boundary of `c`, then `a` boarders `c`.

---

[3] http://www.w3.org/2002/07/owl.

[4] http://swatproject.org/travelOntology.asp.

Below is another snippet of the file about a class:

```
:SuperContinent rdf:type owl:Class ;
    owl:equivalentClass [ rdf:type owl:Class ;
        owl:intersectionOf ( :Island
            [ rdf:type owl:Restriction ;
              owl:onProperty :hasDirectPart ;
              owl:onClass :Continent ;
              owl:minQualifiedCardinality
                  "2"^^xsd:nonNegativeInteger
            ]
        )
    ] .
```

An `owl:Class` called *SuperContinent* is defined here. Interesting features of this class include the following:

1. It is sufficiently and necessarily defined by the `owl:equivalentClass` property, meaning that any instance of a `SuperContinent` should satisfy the description embraced by the equivalent class property, and anything that satisfies such a description is an instance of `SuperContinent`.
2. The description that defines `SuperContinent` belongs to the type called `owl:intersectionOf`, indicating that such a description is a conjunction of two other descriptions. One of them is `Island`, the other, as shown in the code, is an instance of `owl:Restriction`.
3. The `owl:Restriction` specifies its property `hasDirectPart`, its class `Continent` and its cardinality as ≥ 2. This suggests that an instance of this restriction should have the `hasDirectPart` relation to at least two different instances of `Continent`.

Together it defines that a supercontinent is equivalent to an island that has at least two different direct parts that are continents, and vice versa.

In the above example, the OWL annotations provide semantics that can be exploited to uncover hidden information. For example, from the *hasBoundary* relation and *boundaryOf* relation chain between two objects, OWL allows one to infer the hidden *boarders* relation. Nevertheless, due to the sheer level of expressive power offered by OWL 2, such an inference cannot always be performed in finite time without restricting the use of the OWL 2 schema vocabulary. Such restriction can be achieved by composing OWL 2 files with respect to a syntax that corresponds to a computationally decidable logic.

The OWL 2 DL syntax is introduced for this purpose, where DL stands for the Description Logic. The Description Logic (DL) [16] is a family of formal knowledge representations that describe the domain of discourse with concepts (unary predicates), roles (binary predicates) and their instances. DLs have different dialects, which differ from each other on how predicates can be constructed and used. For example, in $\mathcal{SROIQ}$, one of the most expressive and decidable DLs developed so far, a concept expression can be inductively defined as an *atomic concept* A, a singleton *nominal* $\{a\}$, the *top concept* $\top$, the *bottom concept* $\bot$, the *negation* $\neg C$, the *conjunction* $C \sqcap D$, an *existential restriction* $\exists R.C$, a *local reflexivity* $\exists S.Self$ or an *at least restriction* $\geq n\, S.C$, where A is a concept name, a is an individual name, C and D are also concept expressions and R and S are role expressions,[5] which can be either an *atomic role* r or an *inverse role* r^-. With concepts and roles specified, the domain knowledge can be organised by axioms such as *Concept Subsumption* $C \sqsubseteq D$, *Role Chain* $R_1 \circ R_2 \sqsubseteq R_3$, *Class Assertion* $a : C$ and *Role assertion* $(a, b) : R$. A set of such DL axioms is called a DL ontology. In the DL ontology, the set of concept and role axioms is called *TBox*, while the set of assertions is called *ABox*.

OWL 2 DL is underpinned by $\mathcal{SROIQ}$. When an OWL 2 file obeys the OWL 2 DL syntax, it can be regarded as a syntactic variant of an $\mathcal{SROIQ}$ ontology, where classes correspond to concepts and properties correspond to roles. For example, the features of *borders* in Example 1 can be rewritten as the following DL axioms:

$$
\begin{aligned}
\text{Domain Restriction:} &\quad \exists\, borders.\top \sqsubseteq AdministrativeDivision\\
\text{Range Restriction:} &\quad \exists\, borders^-.\top \sqsubseteq AdiminstrativeDivision\\
\text{Role Chain:} &\quad hasBoundary \circ boundaryOf \sqsubseteq borders
\end{aligned}
$$

while the *SuperContinent* definition in Example 1 can be rewritten as a DL axiom as follows:

$$SuperContinent \equiv Island \sqcap \geq 2\ hasDirectPart.Continent$$

Any OWL 2 ontology that does not obey the OWL 2 DL syntactic restriction is said to be an OWL 2 Full ontology.

The logical root of $\mathcal{SROIQ}$ also provides formal semantics to interpret an OWL 2 DL file. Let Δ be the domain, an interpretation function $\cdot^{\mathcal{I}}$ interprets an individual a as a domain entity $a^{\mathcal{I}} \subseteq \Delta$, a concept (class) C as a set $C^{\mathcal{I}} \subseteq \Delta$ and a role (property) R as a set $R^{\mathcal{I}} \subseteq \Delta \times \Delta$. With formal semantics, axioms can also be interpreted. For example, the above axiom specifies that $SuperContinent^{\mathcal{I}}$ is equivalent to the following interpretation:

$$Island^{\mathcal{I}} \cap \{x \mid \#\{y \mid y \in Continent^{\mathcal{I}}, (x, y) \in hasDirectPart^{\mathcal{I}}\} \geq 2\}$$

In other words, *SuperContinent* is the set of domain entities that belong to *Island* and that each has the *hasDirectPart* relation to at least two different domain entities that belong to *Continent*. For more details on the syntax and semantics of $\mathcal{SROIQ}$, we refer interested readers to [120]. Such an $\mathcal{SROIQ}$-based semantics is also called the *Direct Semantics* of OWL 2. OWL 2 Full ontologies, on the other hand, can only be interpreted with the *RDF-based Semantics*.

---

[5] S needs to satisfy a simple role restriction, for which we refer the interested readers to [120].

Table 2.1 Reasoning services

Reasoning services	Explanation
Ontology consistency checking	Checking if an ontology contains contradiction
Classification	Computing the inferrable `OWL:subClassOf` relations between classes
Realization	Computing the inferrable `RDF:type` relation between an individual and a class
Class satisfiablility checking	Checking if a class can have any instance
Axiom entailment checking	Checking if an axiom can be deduced from an ontology
Conjunctive query answering	Answering a query against an ontology

With formal semantics and decidable logic underpinning, the ontology reasoning can be supported by automated reasoners. In an ontology, the typical reasoning services are illustrated in Table 2.1.

These reasoning services have been implemented by numerous reasoners such as FaCT++ [233], Pellet [216], HermiT [213], TrOWL [231] and Knoclude [221].

Although OWL 2 DL is computationally decidable, it is still rather complex to reason with. To address this issue, the designers of OWL 2 further restrict its syntax to develop three OWL 2 profiles for which most of the reasoning services, except conjunctive query answering, can be performed in polynomial time. These profiles are as follows:

1. **OWL 2 EL** is an OWL 2 profile designed to offer an efficient classification of large terminologies. It is based on the $\mathcal{EL}$ family of DLs and supports expressive powers such as class intersections (`owl:intersectionOf`), qualified existential restrictions (`owl:someValuesFrom`) and property chains (`owl:propertyChainAxiom`). OWL 2 EL has been widely used in many of the largest ontologies developed so far, e.g. the SNOMED CT [219] ontologies.
2. **OWL 2 QL** is an OWL 2 profile designed to offer efficient query answering services over a large amount of data. It is based on the DL-Lite family of DLs and supports expressive powers such as inverse properties (`owl:inverseOf`). Notably, conjunctive query answering of an OWL 2 QL ontology can be reduced to SQL query answering in a relational database, which makes it possible to enjoy the systems and optimisations which have been developed for a relational database. This relation with the relational database also makes OWL 2 QL an ideal candidate for the semantic upgrade of traditional database data and ontology-based data access.
3. **OWL 2 RL** is an OWL 2 profile inspired by Description Logic Programs. It supports features such as functionality (`owl:FunctionalProperty`), and its

reasoning mechanism can be implemented using a rule-based reasoning engine. Therefore it is suitable for scenarios with deductive database and Datalog engines. For example, the Oracle Spatial and Graph Database[6] implements a built-in OWL 2 RL reasoner.

The *OWL 2 Web Ontology Language* (OWL 2) [171] is the state-of-the-art ontology language for the Semantic Web with formal semantics. Currently, it enjoys the W3C Recommendation status. *OWL 2* is an extension of *OWL*, which was the most famous ontology language. It aims to extend the expressiveness of the *OWL* specification by introducing new constructs.

This section presents only a very brief introduction of OWL 2. For more interested readers, we refer to the OWL 2 Overview.[7]

2.1.3 SPARQL

Now that we have RDF and OWL for constructing knowledge graphs and their schemas, we will introduce the standard query language for RDF and OWL.

Overview and Background

The SPARQL Protocol And RDF Query Language is a query language tailored to retrieve and manipulate data within RDF graphs. It is one of the core technologies underlying the Semantic Web paradigm. Built on earlier RDF query languages such as rdfDB, RDQL and SeRQL, **SPARQL 1.0** became an official W3C Recommendation on 15 January 2008. It was standardised by the *RDF Data Access Working Group* (RDAWG) as part of the *W3C Semantic Web Activity*. RDAWG defined the syntax and semantics of the query language and later on extended the SPARQL technology to include some of the features that the community has identified as both desirable and important for interoperability based on the experience with the standard's initial version.[8] On 21 March 2013, 11 recommendations specifying the actual **SPARQL 1.1** version were released.[9] Improvements with respect to the first version included RDF graph update support, a more powerful query language supporting sub-queries, aggregate operators like `count` and a simplified negation form, and a definition of serialisation formats.

---

[6]http://www.oracle.com/technetwork/database/options/spatialandgraph/overview/index.html.

[7]http://www.w3.org/TR/owl2-overview/.

[8]http://www.w3.org/2011/05/sparql-charter.

[9]http://www.w3.org/TR/sparql11-overview/.

In the following, the most important characteristics of **SPARQL 1.1** will be introduced and some illustrative examples will be provided in order to let the reader acquire an essential understanding of the query language. For a complete reference, please refer to the W3C recommendations.

SPARQL Queries

SPARQL statements are expressed according to the Turtle syntax,[10] smoother than XML, and are based on the *pattern-matching* mechanism. The basic fragment of an SPARQL query resembles an RDF triple (subject, predicate, object) as in the following example[11] in which variables occur in the subject and object positions and `dbpedia-owl` is the prefix for http://dbpedia.org/ontology/.

```
?movie dbpedia-owl:country ?country
```

This fragment will match RDF triples related to the DBpedia ObjectPropery `country` by substituting the variables. A simple query has a structure similar to that of SQL and looks like the following.

```
PREFIX dbpedia-owl: <http://dbpedia.org/ontology/>
SELECT ?movie, ?director
WHERE {
  ?movie a dbpedia-owl:Film.
  ?movie dbpedia-owl:director ?director.
  ?director dbpedia-owl:birthPlace <http://dbpedia.org/
      resource/Italy>
} LIMIT 100
```

The simple query mentioned above retrieves the first 100 movies having an Italian director. In the example, ORDER BY ASC(?director) could be used to order results and OFFSET 10 may be added to skip initial items. Differently from SQL the FROM clause is optional, which can be used to specify the default RDF graph or the dataset to be used for matching. SPARQL also allows more complex queries which may include union, optional query parts, filters, value aggregation, path expressions, nested queries, etc.

---

[10]http://www.w3.org/TeamSubmission/turtle/.

[11]DPpedia http://dbpedia.org/ has been used to provide supporting examples.

```
PREFIX dbpedia-owl: <http://dbpedia.org/ontology/>
SELECT ?movie, ?director, ?place, ?composer
WHERE {
  ?movie a dbpedia-owl:Film.
  ?movie dbpedia-owl:director ?director.
  ?director dbpedia-owl:birthPlace ?place.
OPTIONAL {?movie dbpedia-owl:musicComposer ?composer}
FILTER (?place = <http://dbpedia.org/resource/Italy>  or ?
    place = <http://dbpedia.org/resource/Spain> )
} OFFSET 1 LIMIT 3
```

In the above example, a complex query is shown. The query looks for all movies made by either an Italian or a Spanish director together with their soundtrack composer, if any, and returns only three results skipping after the first. The OPTIONAL keyword is used to include in the results also the movies whose soundtrack composer is not known; while the FILTER keyword allows to add constraints. The following table lists the result obtained by querying the DBpedia SPARQL endpoint.[12]

```
prefix http://dbpedia.org/resource/
movie director  place composer
dbp-res:Io_sto_con_gli_ippopotami dbp-res:Italo_Zingarelli
     dbp-res:Italy dbp-res:Walter_Rizzati
dbp-res:My_Name_Is_Janez_Jan%C5%A1a dbp-res:Janez_Jan%C5%
    A1a_(performance_artist) dbp-res:Italy
dbp-res:Los_peces_rojos dbp-res:Jos%C3%
    A9_Antonio_Nieves_Conde  dbp-res:Spain dbp-res:
    Miguel_Asins_Arb%C3%B3
```

Although as a query language SPARQL only retrieves information explicitly defined in the model without committing to performing any inference, it could be used together with ontological information in the form of, for example, RDF Schema or OWL axioms. The *SPARQL 1.1 Entailment Regimes* specification defines which answers should be given under which entailment regime, specifying entailment regimes for RDF, RDF Schema, D-Entailment, OWL and RIF.

Apart from SELECT queries, SPARQL supports ASK queries that provide boolean answers, and CONSTRUCT queries. The latter whose result is used to build new RDF graphs can be constructed from a query result. Any construct used in SELECT queries can be used in both ASK and CONSTRUCT queries.

[12]http://dbpedia.org/sparql:

SPARQL Updates

SPARQL Update operations allow to create, update and remove RDF graphs. Those operations are performed against a Graph Store and can be grouped into sequences. For instance, the following request inserts a triple into the default graph asserting that Julie was born in England.

```
PREFIX dbpedia-owl: <http://dbpedia.org/ontology/> .
INSERT DATA \{ <http://www.sparql.example/person#Julie>
    dbpedia-owl:birthPlace <http://dbpedia.org/resource/
    England>. \}
```

As in other query languages (e.g. SQL) insertions and deletions may be dependent on query results. The example given below shows how to query

```
PREFIX dbpedia-owl: <http://dbpedia.org/ontology/> .
INSERT { ?person dbpedia-owl:stateOfOrigin <http://dbpedia.
    org/resource/England> }
WHERE
  { ?person dbpedia-owl:birthPlace <http://dbpedia.org/
      resource/London>
  }
```

In both queries the default graph is used, while the WITH keyword can be used to choose a specific graph to be updated.

SPARQL also includes shortcut operations in order to operate at the graph level. Those operations include: CREATE for creating a new graph; DROP for removing the specified graph(s); CLEAR for removing all the triples in the specified graph(s); COPY for inserting all data from an input graph into a destination graph; MOVE for moving all data from an input graph into a destination graph; ADD for inserting all data from an input graph into a destination graph; LOAD for reading an RDF document from an IRI and inserting its triples into the specified graphs.

It should be noted that, in general, answering SPARQL queries goes beyond simple graph matching, some kind of reasoning support is usually needed, in order to account, e.g. the schema and the *owl:sameAs* links.

2.2 Ontologies and Vocabularies

A key element of a knowledge graph is the ontology it has as its schema. In a nutshell, the ontology allows to describe and represent all the information present within the knowledge graph.

2.2.1 *Some Standard Vocabularies*

In order to speed up the development process of knowledge graphs and save resources, we need to reuse the already existing vocabularies as much as possible. The underlying idea of adopting existing vocabularies is to enable optimal reuse of the work that has already been done. In this sense, it is important to first take the time to look at what is currently available. Therefore in this section we present an overview of the most "well-known vocabularies" for representing the information in a knowledge graph.

Organisation Ontology

The organisation ontology supports the publishing of organisational information across a number of domains [201]. The ontology provides terms to support the representation of

- organisational structure that covers: (1) the notion of an organisation, (2) decomposition into suborganisations and units, and (3) purpose and classification of organisations
- reporting structure that covers: (1) membership and reporting structure within an organisation, and (2) roles, posts and relationship between people and organisations
- local information, which includes sites or buildings, locations within sites
- organisational history, which includes merger and renaming

Figure 2.8 depicts the main components of the ontology.

GoodRelations

The GoodRelations is an ontology for e-commerce that defines terms for describing products, price and company data [114]. and for defining a data structure that is:

- industry neutral, which means, it is suited for consumer electronics, cars, tickets, real estate, labour, services or any other type of goods,
- valid across the different stages of the value chain, which implies, from raw materials through retail to after-sales services and
- syntax neutral, which means, it should work in RDFa, Turtle, JSON or any other popular syntax.

The main components of the ontology for representing e-commerce scenarios are

- an **agent** that can be a person or an organisation
- an **object** that can be a camcorder, house, car, etc.; or **service**, e.g. a haircut.

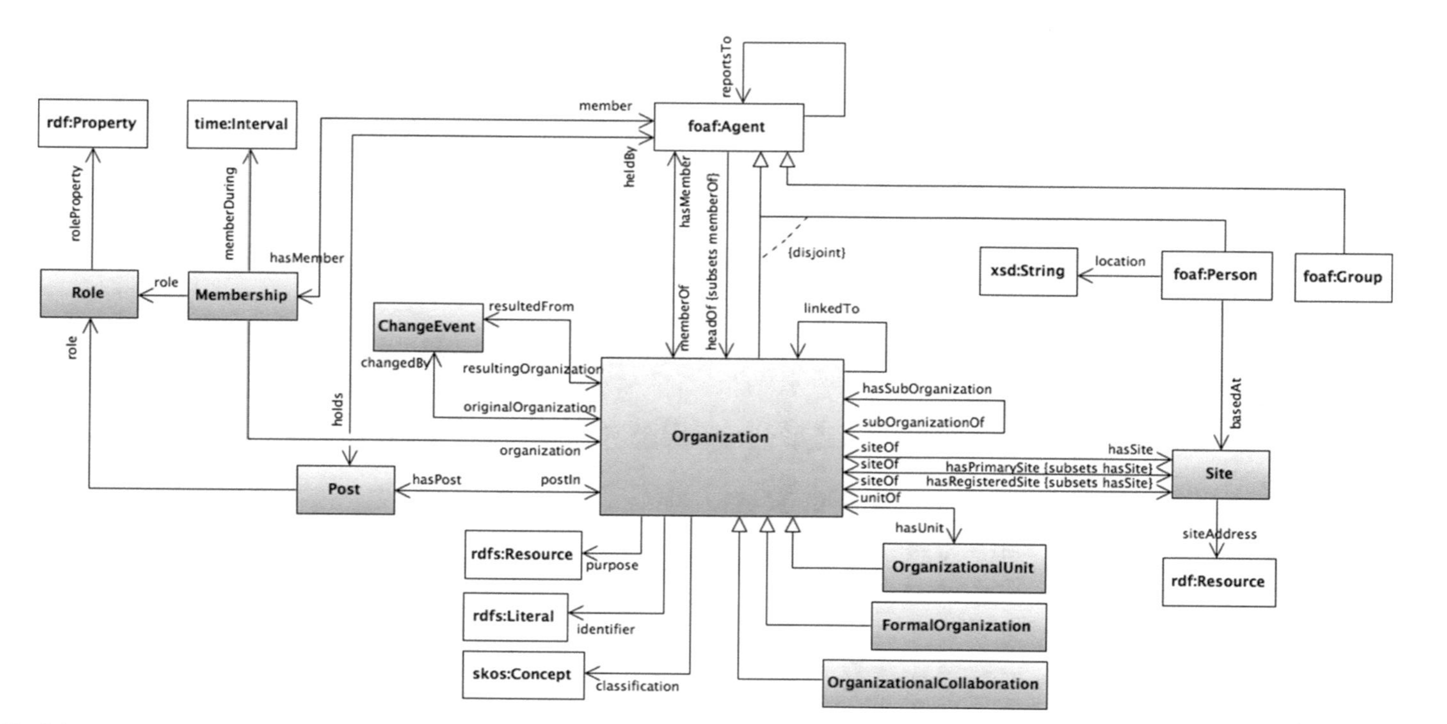

Fig. 2.8 Organisation ontology [201]

- a **promise** or an offer, to transfer some rights on the object or to provide the service for a certain compensation, made by the agent and related to the object or service,
- a **location** from which the offer is available.

GoodRelations [114] is a generic model of information for offering any kind of goods to others and for specifying the expected compensation and conditions. Figure 2.9 describes the main components of the ontology.

Data Cube Vocabulary

The Data Cube Vocabulary provides a means to publish multidimensional data, such as statistics, on the Web. The model underpinning the Data Cube vocabulary is compatible with the cube model that underlies SDMX (Statistical Data and Metadata eXchange) [96], an ISO standard for exchanging and sharing statistical data and metadata among organisations. The Data Cube Vocabulary in turn builds upon the following RDF vocabularies: (1) SKOS [124] for concept schemes, SCOVO [107] for core statistical structures, Dublin Core Terms [158] for metadata, VoiD [7] for data access, FOAF [39] for agents and ORG [201] for organisations.

Within the Data Cube Vocabulary, a DataSet is a collection of statistical data that corresponds to a defined structure. This data can be described as one of the following kinds:

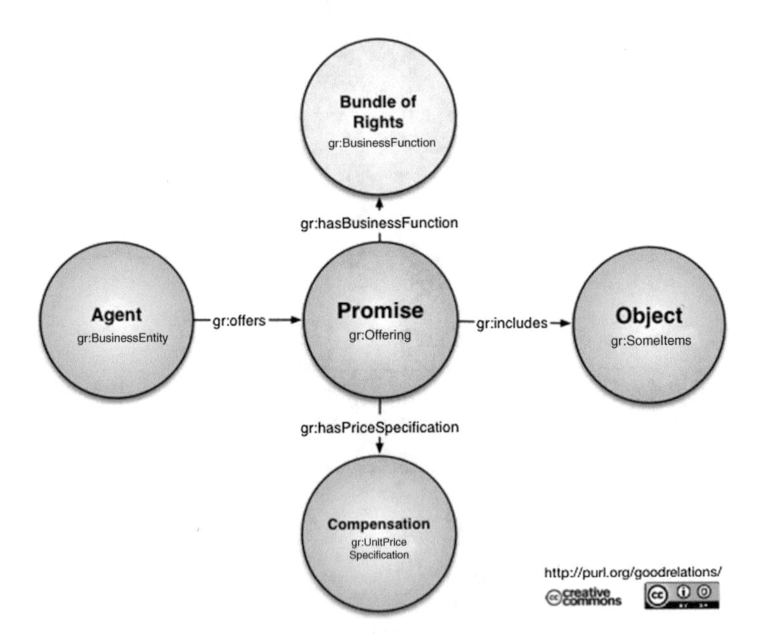

Fig. 2.9 GoodRelations ontology [114]

- Observations, which are the actual data, are the measured values. In a statistical table, observations are the values in the table cells.
- Organisational structure, which describes the values of each dimension at which the observations are located.
- Structural metadata, to be able to interpret the dataset, for example we need to know the unit of measurement, or if it is a normal value or a series break. These metadata are provided as attributes and can be attached to individual observations or to higher levels.
- Reference metadata, which describes the dataset as a whole, such as categorisation of the dataset, its publisher, etc.

Moreover, the main components of a cube are:

- *dimensions* which are used to identify the observations; a set of values for all the dimension components is sufficient to identify a single observation
- *measure* which represents the phenomenon being observed
- *attributes* that allow us to qualify and interpret the observed values they enable the specification of the units of measure, any scaling factors and metadata.

Figure 2.10 depicts the main core components of the RDF Data Cube Vocabulary.

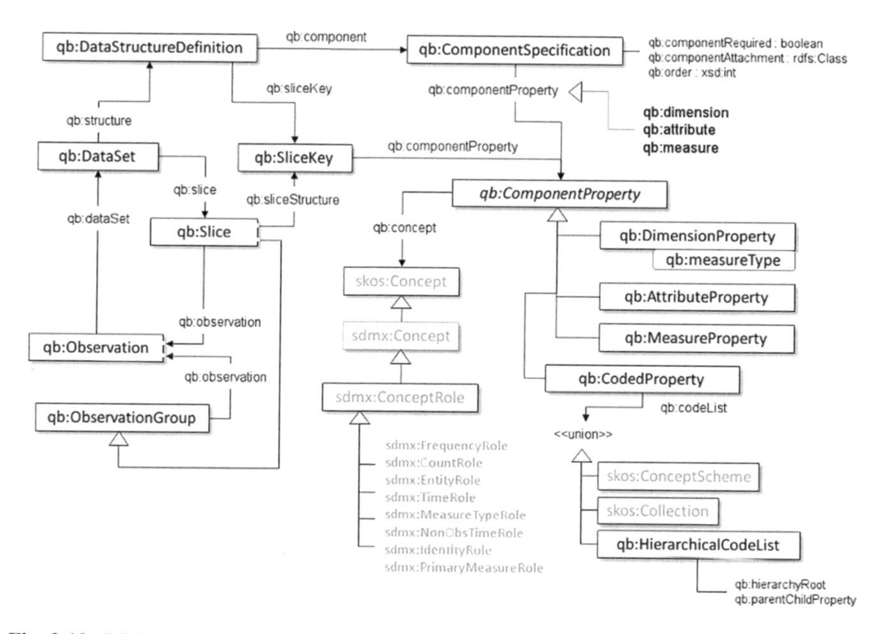

Fig. 2.10 RDF Data Cube Vocabulary core components [60]

Friend-of-a-Friend (FOAF)

The Friend-of-a-Friend (FOAF) ontology defines terms for describing people, their activities (collaboration) and their relations to other people and objects [39]. Different kinds of applications can use or ignore different parts of FOAF. We can group the FOAF terms in the following categories:

- *Core*, which includes classes for describing the characteristics of people and social groups that are independent of time and technology. Moreover, FOAF defines classes for *Project*, *organisation* and *Group* as other kinds of agents.
- *Social Web*, which includes classes for describing an Internet account, address books and other Web-based activities.
- *Linked Data Utilities*, which include a few "demonstration" terms that served largely educational purposes, e.g. geekcode, alongside a few technical utility terms, e.g. focus, LabelProperty, that support wider information-linking efforts.

2.2.2 *schema.org*

In early June 2011, big players such as Google, Yahoo! and Bing introduced *schema.org*, a collection of terms that can be used to mark up HTML pages to improve the display of search results. This shared markup vocabulary makes it easier for Web masters to decide on a markup schema and get the maximum benefit for their efforts.

The data model used is very generic and derived from RDF Schema. *schema.org* data model has a set of types, arranged in a multiple inheritance hierarchy where each type may be a subclass of multiple types. Moreover, there are a set of properties where;

- Each property may have one or more types as its domains. The property may be used for instances of any of these types.
- Each property may have one or more types as its ranges. The value(s) of the property should be instances of at least one of these types.

The canonical machine representation of schema.org is in RDFa and is available here http://schema.org/docs/schema_org_rdfa.html. The type hierarchy of *schema.org* is not intended to be a "global ontology" of the world. It only covers the types of entities for which Microsoft, Yahoo!, Google and Yandex, think they can provide some special treatment for, through their search engines, in the near future. The most generic type is *Thing*, and the most commonly used types are:

- Creative works: CreativeWork, Book, Movie, MusicRecording, Recipe, etc.
- Embedded non-text objects: AudioObject, ImageObject, VideoObject
- Event
- Health and medical types
- Organisation

- Person
- Place, LocalBusiness, Restaurant
- Product, Offer, AggregateOffer
- Review, AggregateRating
- Action

In this section, we presented an overview of most of the "well-known vocabularies” for representing the information within large organisations in a knowledge graph.

2.3 Data Lifting Standards

In many large organisations, the data or knowledge might take various formats, such as relational databases, Web pages, documentations, transaction logs, etc. To make these information accessible in the organisation’s knowledge graph, it requires to convert them from their current representation into the format of knowledge representation. In our scenario, it is the RDF data model. The conversion process is called *data lifting*, which means the conversion is not only a transform from one format to another, but also a “lift” of the information from the data level into the machine-readable “knowledge” level.

There are various approaches available to perform the data lifting. For example, to extract knowledge from natural language texts or Web pages, there are approaches of named entity recognition, information extraction, concept mining, text mining, etc. There are many tools or libraries available in either open source or commercial licenses, such as GATE,[13] OpenNLP[14] and RapidMine.[15]

In this section, we lay special focus on two W3C standards on data lifting, i.e. RDB2RDF and GRDDL. These two standards cover the data lifting from structured or semistructured legacy data and are probably the most important data formats in large organisations. RDB2RDF specifies how to translate relational data into the RDF format (introduced in Sect. 2.3.1) and GRDDL defines the standard approaches to translate the XML data into RDF (briefly presented in Sect. 2.3.2).

2.3.1 RDB2RDF

If you are in the middle of constructing knowledge graphs for large organisations, or maybe you only want to enjoy the benefits of Linked Open data (e.g. the mature ecosystems of reasoning or data integration), most likely you will have to convert

[13]https://gate.ac.uk/.

[14]http://opennlp.apache.org/index.html.

[15]https://rapidminer.com/.

Table 2.2 RDB2RDF: direct mapping or R2RML

Considerations	Direct mapping	R2RML
Automated mapping	Y	N
Customisable vocabulary	N	Y
Customisable URI	N	Y
Extraction-transform-load	N	Y
Example scenarios	LOD Publish e.g. SPARQL endpoint, entity data publishing (HTTP303)	Data Integration e.g. reuse popular ontologies/vocabularies, Reasoning Service e.g. Consistency checking, Deriving new knowledge

your legacy data from a relational database into a Linked Data format, i.e. RDF or OWL. If this is the case for you, we have good news, that is, you don't have to start from scratch because there is a "standard" way to do that. Even better, you can also find useful tools to speed up your work. In this subsection, we will introduce the W3C recommendations for translating your relational data into an RDF format and the list of tools which are useful for this task.

In 2012, the RDB2RDF W3C working group published two recommendations to standardise languages for mapping relational data and relational database schema into RDF and OWL. These two recommendations are designed for two typical scenarios of converting relational data into RDF data. The first recommendation is called "*A Direct Mapping of Relational Data to RDF*".[16] If you prefer a quick conversion and your database schema is designed to be good enough (e.g. well-defined primary keys and foreign keys, meaningful table and column names, etc.), then direct mapping can be a good choice. The only input in this case is the database (data and schema) and the output is the RDF version of your data. It is simple but you don't have much control over the conversion settings. The second recommendation is "*R2RML: RDB to RDF Mapping Language*".[17] Using an R2RML, you can customise the mappings to generate the RDF data based on your *design*. For example, if you would like to generate your RDF data by reusing some popular vocabularies or your predefined domain ontologies, you will go for R2RML. Table 2.2 gives some of the possible considerations for making a choice between the two specifications.

Direct Mapping

When you want to play with the RDF version of your data and do not want to bother learning the R2RML language, direct mapping is the one you might start with for RDFising your data. Instead of going into the details of specification, we illustrate the main conversion details by a simple example. Suppose your data is about the project development which contains information about projects, developers and task assignments. The input for direct mapping is only the database (including both data

[16]http://www.w3.org/TR/rdb-direct-mapping/.

[17]http://www.w3.org/TR/r2rml/.

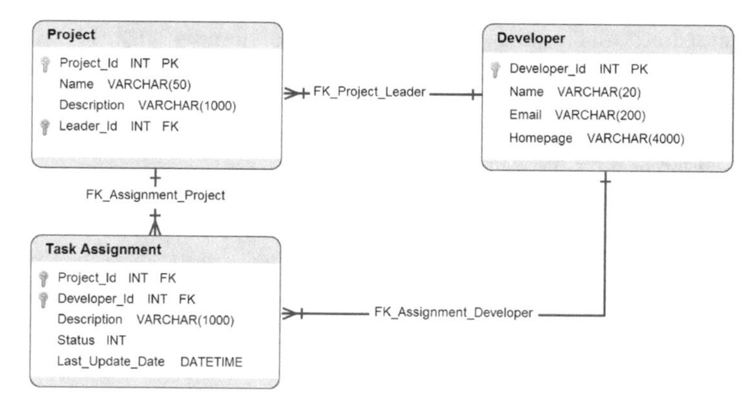

Project_Id	Name	Description	Leader_Id
6	K-Drive	K-Drive is an EU IAPP...	12
7	Whatif	Whatif is a project funded by...	12

Developer_Id	Name	Email	Homepage
12	Jeff	jeff.z.pan@...	NULL
16	Boris	boris@...	NULL

Project_Id	Developer_Id	Descriptino	Status	Last_Updated_Date
6	12	Research Fellow Recruitment	0	2014-12-16
6	16	User Interface Prototype	1	2014-08-12

Fig. 2.11 Direct mapping example: project development

and schema definitions) and the output is the translated RDF data. A sample database of our example might look like the one shown in Fig. 2.11. The upper part is the schema, i.e. definitions of three tables and primary/foreign keys, and the lower part is the data, i.e. data rows in three tables. Essentially, the direct mapping specification is an algorithm to carry out the automatic conversion. We now introduce the main steps of the process on the sample database.

In the direct mapping, the conversion is processed in an entity-centric way. Specifically, each data row is viewed as a set of triples describing an entity. Hence, the first step is to identify the entity (an RDF resource) from each data row. Depending on the existence of a primary key, two situations will be considered when generating RDF resources for data rows.

When the table has a primary key, a URI resource will be generated for each row in this table. Taking the *Project* table for example (cf. Fig. 2.11), there is a primary key defined on column *Project_Id*. Given a predefined URI base http://abc.org/DB/, a URI resource of http://abc.org/DB/Project/Project_Id=6 will be generated for the first row (*Project_Id* = 6) of the *Project* table. The syntax to be followed when generating a URI resource takes the form of $'URI\_BASE' + ' COLUMN1\_NAME = COLUMN1\_VALUE; COLUMN2\_NAME = COLUMN2\_VALUE...'$, where URI_BASE is the URI prefix and $\{COLUMN1, COLUMN2, ...\}$ is the set of columns in the primary key definition of the table.

The second situation is for the tables without primary key definitions. In this case, a blank node will be defined for each table row. For example, the table *Task_Assignment* does not have a primary key. For its first row, a blank node of _:*b*5 might be generated as the internal ID of the centric entity. No particular syntax is defined for generating a blank node. As long as the node ID is unique, it is a valid one.

After generating the RDF resource for a data row, the next step is to convert the data row into RDF triples describing the newly generated resource. The very first triple to be generated is the type assertion. This triple simply specifies that the data row resource is an instance of its table class (a class generated for the table). For example, we can have the type assertion of *Project* table's first row as $< DB{:}Project/Project\_Id = 6, rdf{:}type, DB{:}Project >$, where *DB* denotes the aforementioned base URI.

In addition to the type assertion, all column values of the data row need to be converted into triples. Depending on the table schema definitions, two types of triples can generated. The first type is data-valued triples, which have literal values as the objects. They are generated from columns that are not involved in any foreign key definition.

For those foreign key columns, the generated triples will be relational, which means that their objects are either URI resources or blank nodes. Looking at the *Project* table schema, there is no foreign key defined on the *name* column. Hence, the *name* value of its K-Drive data row will be generated as a triple $<DB{:}Project/Project\_Id = 6$,
DB:*Project#Name*, "K-Drive".

The other type of columns is the foreign key columns. For example, the *Leader_Id* column is specified as a foreign key referencing the *Developer_Id* in the *Developer* table. This column value will be converted into a relation between the K-Drive project and its leader developer, i.e. Jeff. The main aspect to be considered in this conversion is how to get the RDF resource of the object in the triple. Given the fact that a foreign key is referencing the other data row (which is usually in another table), the entity denoted by the foreign key column(s) should be generated from the referenced data row accordingly. On finding the referenced data row, the RDF resource can be generated by the same logic we introduced in the *GETRowRES* function. In our example, the entity denoted by *Leader_Id* value 12 needs to be generated from the first row of the *Developer* table, which will be $DB{:}Developer/Developer\_Id = 12$. Eventually, the triple to be converted from the *Leader_Id* column is $< DB{:}Project/Project\_Id = 6, DB{:}Project\#ref\text{-}Leader\_Id, DB{:}Developer/Developer\_Id = 12 >$.

By applying the above-mentioned logics on all data tables, the final RDF conversion of our sample database is similar to the one cited in Listing 2.1.

Listing 2.1 RDF Version of Project Development RDB

```
@base <http://abc.org/DB/>
@prefix xsd: <http://www.w3.org/2001/XMLSchema#> .
<Project/Project_Id=6> rdf:type <Project> .
<Project/Project_Id=6> <Project#Project_Id> 6.
<Project/Project_Id=6> <Project#Name> "K-Drive".
<Project/Project_Id=6> <Project#Description> "K-Drive is an
     EU IAPP...".
<Project/Project_Id=6> <Project#ref-Leader_Id> <Developer/
    Developer_Id=12>.
<Project/Project_Id=7> rdf:type <Project> .
<Project/Project_Id=7> <Project#Project_Id> 7.
<Project/Project_Id=7> <Project#Name> "Whatif".
<Project/Project_Id=7> <Project#Description> "Whatif is a
    project funded by...".
<Project/Project_Id=7> <Project#ref-Leader_Id> <Developer/
    Developer_Id=16>.

<Developer/Developer_Id=12> rdf:type <Developer> .
<Developer/Developer_Id=12> <Project#Developer_Id> 12.
<Developer/Developer_Id=12> <Project#Name> "Jeff".
<Developer/Developer_Id=12> <Project#Email> "jeff.z.pan@
    ...".
<Developer/Developer_Id=16> rdf:type <Developer> .
<Developer/Developer_Id=16> <Project#Developer_Id> 16.
<Developer/Developer_Id=16> <Project#Name> "Boris".
<Developer/Developer_Id=16> <Project#Email> "boris@...".

_:b5 rdf:type <Task_Assignment> .
_:b5 <Task_Assignment#ref-Project_Id> <Project/Project_Id
    =6> .
_:b5 <Task_Assignment#ref-Developer_Id> <Developer/
    Developer_Id=12> .
_:b5 <Task_Assignment#Status> 0 .
_:b5 <Task_Assignment#Last_Updated_Date> "2014-12-16" .
_:b6 rdf:type <Task_Assignment> .
_:b6 <Task_Assignment#ref-Project_Id> <Project/Project_Id
    =7> .
_:b6 <Task_Assignment#ref-Developer_Id> <Developer/
    Developer_Id=16> .
_:b6 <Task_Assignment#Status> 1 .
_:b6 <Task_Assignment#Last_Updated_Date> "2014-08-12" .
```

RDB2RDF Mapping Language

As already shown, direct mapping is a very efficient way to achieve a quick RDF conversion of your RDB data. However, in many real-world scenarios, this direct mapping might not be sufficient. For example, when converting the project database into RDF, one might want to use popular domain ontologies like DOAP (Description of a Project Vocabulary[18]). The motivation is that using a popular domain ontology

[18]http://usefulinc.com/ns/doap.

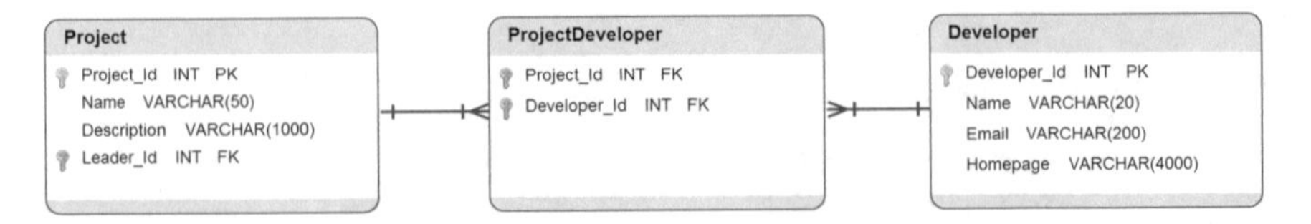

Fig. 2.12 Example: project development (more tables)

improves the visibility of your knowledge base and makes it much easier to be integrated or linked by other knowledge bases. Furthermore, in some situations, a direct mapping might generate unwanted conversions. For example, suppose that there is one more table that simply records which developers are involved in which projects as shown in Fig. 2.12. A direct mapping will create blank nodes to represent such relations, which is obviously not the most efficient way. A more efficient and simpler mapping might be to use a property such as *involvedIn* to directly specify the relation between developers and projects. In some situations, you simply do not want to share some part of your data due to confidential or security considerations. For example, you might want to prevent publishing the emails of developers in the RDF version of your knowledge base.

All the above three examples require a customised conversion which a direct mapping cannot provide. The first example requires *customised RDF resource generation*, the second example needs to *customise the mapping* and the third example requires the ability to *extract part of the data for conversion*. All these requirements correspond to the main constructs of the RDB2RDF Mapping Language as follows:

- *Term Maps* In the RDB2RDF specification, the RDF terms are used to denote all types of RDF resources of IRI, blank nodes and literals. The term map is essentially a function which is capable of generating customised terms from data rows, which makes a domain ontology reuse possible.
- *Logical Table* A logical table, as its name indicates, is a virtual table which is constructed from "real" tables. This table enables customised data extraction before doing the triple conversion, which meets the requirement of our third example.
- *Triple Maps* This map mechanism is designed for the ability to specify how triples are generated from data rows. It enables customised mapping, which solves issues in our second example.

In the rest of this subsection, we will go through the three constructs.

Term Map

A term map is a function to generate an RDF resource from data rows (either those from database definitions such as tables or views, or the logical ones which we will introduce very soon). For the *Project* table in our example database, suppose that we would like to generate a type assertion for each data row. Instead of the "strange" vocabulary URIs to be generated by a direct mapping, we would prefer to use the DOAP vocabulary, e.g. using *doap:Project*[19] as the class name. To achieve this feature in R2RML, we can use the following mapping definition.

```
[] rr:predicateMap [ rr:constant rdf:type ];
   rr:objectMap [ rr:constant doap:Project ].
```

Note that no matter whatever the content of the data row, the mapping will generate the same pairs of RDF resources: *rdf:type* and *doap:Project* for the predicate and object components of the generated assertion accordingly. This is the so-called constant-valued term map.

In addition to constant resources, the most common task in data transformation is to convert the data values from the table to RDF resources. If you simply want to convert the value of a data column into a literal resource, in R2RML, the simplest way to do that is to use a column-valued term map. If defined in our sample *Project* table, the following mapping will generate a literal resource as the object of a triple using the value of the *name* column. Note that the values followed by an *rr:column* should be valid column names.

```
[] rr:objectMap [ rr:column "Name" ]
```

The most customisable term map is the last type of *template-valued* term map. One can design his/her own URI scheme using a string template. The syntax is quite straightforward. Put down the constant strings. Wherever you want to have a column value to be part of it, simply put the column name in the right position and surround it with curly brackets. The following example defines our customised project URIs using the *Project_Id* column as the variable part.

```
[] rr:subjectMap [ rr:template "http://abc.org/DB/Project/
    ID/{Project_Id}" ].
```

Logical Table

In R2RML, the logical table is a way to enable the customised data extraction and transformation from the original database using SQL queries. For example, in the *Developer* table, if one would like to omit the email information in its RDF ver-

[19]xmlns:doap="http://usefulinc.com/ns/doap#".

sion, the following logical table can be defined using a simple SQL query to select necessary columns only.

```
[] rr:sqlQuery """
        Select Developer_Id,
        Name,
        Homepage
          from Developer
   """.
```

In addition to supporting the data extraction, using SQL queries makes it possible to carry out data transformations using the database's built-in functions, e.g. MD5 is a built-in function of MySQL since version 5.5.3. For example, one would like to use developer's emails as their unique IDs because the *Developer_Id* is not globally unique. The following logical table might make more sense in the open data environment.

```
[] rr:sqlQuery """
        Select MD5(Email) as Developer_Id,
        Name,
        Homepage
          from Developer
   """.
```

Of course, the simplest form of logical tables is by directly reusing tables or views defined in the databases. For example, the *Developer* table can be simply referenced like *[] rr:tableName "Developer"*.

Triple Maps

Both term maps and logical tables are fractions or components of the mapping definition. To put them together to define the transformation, we need the triple maps. A triple map specifies how a data row is converted into a set of RDF triples. The following triple map defines the rule to convert the *Name* and *Description* columns of a *Project* table into two RDF triples.

```
[]
    rr:logicalTable [ rr:tableName "Project" ];
    rr:subjectMap [  rr:template "http://abc.org/DB/Project
       /ID/{Project_Id}" ];
    rr:predicateObjectMap [
        rr:predicateMap [rr:constant DB:name];
        rr:objectMap [ rr:column "Name" ];
    ];
    rr:predicateObjectMap [
        rr:predicate DB:decription;
        rr:objectMap [ rr:column "Description" ];
    ].
```

Each triple map has one logical table and one subjectMap. Based on a term map, the subjectMap defines the subject resource of the triple(s) to be generated on each data row. There can be one or more predicateObjectMaps, which define the pair(s) of predicate and object resources of triples. In our examples, there are two predicateObjectMaps, each of which defines a triple to be generated on a column-valued term map. Note that in the second predicateObjectMap, a simpler syntax of predicateMap is used. It is called a constant shortcut property, where the *rr:constant* can be omitted using an *rr:predicate* directly followed by the constant resource. Based on the sample mapping, the triples to be generated are as follows.

Listing 2.2 Triple Map Result

```
@base <http://abc.org/DB/>
@prefix DB: <http://abc.org/DB/>.
<Project/ID/6> DB:name "K-Drive".
<Project/ID/6> DB:desription "K-Drive is an EU IAPP...".
<Project/ID/7> DB:name "Whatif".
<Project/ID/7> DB:description "Whatif is a project funded
    by...".
```

Going back to one of our motivation examples of Fig. 2.12, the following triple map can be used to generate the relational triples between projects and developers in a concise way.

```
[]
    rr:logicalTable [ rr:tableName "ProjectDeveloper" ];
    rr:subjectMap [  rr:template "http://abc.org/DB/Project
        /ID/{Project_Id}" ];
    rr:predicateObjectMap [
        rr:predicate DB:hasDeveloper;
        rr:objectMap [ rr:template "http://abc.org/DB/
            Project/ID/{Developer_Id}" ];
    ];
```

RDB2RDF Tools

There are various implementations of an RDB2RDF specification. The W3C published a report on the implementations in 2012. Although it is slightly outdated, it is still worth a reference: http://www.w3.org/TR/rdb2rdf-implementations/. As a complement to the W3C's list, we briefly introduce some of the most well-known tools as follows:

- Morph-RDB (formerly called ODEMapster) is an RDB2RDF engine developed by the Ontology Engineering Group. It supports two operational modes: data upgrade (generating RDF instances from data in a relational database) and query translation (SPARQL to SQL). Available at: https://github.com/oeg-upm/morph-rdb

- The D2RQ Platform is a system for accessing relational databases as virtual, read-only RDF graphs. It offers an RDF-based access to the relational database content without having to replicate it into an RDF store. Available at: http://d2rq.org/
- Some triple stores or RDF database systems also have their own implementations of RDB2RDF. For example, Virtuoso[20] has its own previously developed proprietary equivalent of R2RML called Linked Data Views. Now, Virtuoso supports R2RML by the inclusion of a simple translator which basically translates R2RML syntax to Virtuoso's own Linked Data Views syntax. See also: http://virtuoso.openlinksw.com/dataspace/doc/dav/wiki/Main/VirtR2RML.

2.3.2 GRDDL

In addition to residing in relational databases, the legacy data to lift might also be stored in other formats. Among others, a common one could be the XML format. For example, the purchase orders from the retailer section are encoded in an XML format, the data captured by sensors uses XML syntax or the information in question is simply published as Web pages (XHTML) on the Web. Although all these data are stored in XML, their syntaxes and semantics are potentially totally different, which impedes their integration to the organisation's knowledge graph.

To make use of all these various XML data sources in your organisation's knowledge graph, you might want to convert them into an RDF representation. Essentially, this is a process to transform the data from one XML format to another. Although such a transformation is not a complicated task, there are many factors to be considered in the process, e.g. how to specify whether or not a transformation has been provided for the XML resource; if so, what is the transformation algorithm and how to specify the algorithm's location. Different approaches can be provided for each of these questions. Without a unified pipeline, the heterogeneity of the solutions might cause another obstacle in knowledge integration. The good news is that there is a W3C recommendation standard, GRDDL,[21] which is dedicated to provide a standard way for deriving resource descriptions (in RDF) from dialects of XML languages.

Essentially, what GRDDL specifies is a standardised way to declare whether an XML document contains information compatible with RDF and (if so) where is/are the transformation algorithm(s). Although there are many solutions to implement such transformation algorithms, the very common approach is using one of the W3C's standards, i.e. XSLT—the transforming language for XML. GRDDL standardises the above specification for both general-purpose XML documents and XHTML Web pages. In addition, it also allows to specify the gleanable data in "*meta-documents*" of XML namespace documents and XHTML profiles, which will result in such specifications being applied to every document associated with the meta-documents.

[20] http://virtuoso.openlinksw.com/.

[21] http://www.w3.org/TR/grddl/.

In this subsection, we will go through an example based on an XHTML Web page about a product using the GoodRelations Vocabulary.

Product Web page Example

Suppose that Company A has crawled a large number of Web pages about the products from its competitor, Company B. To integrate the product information into its knowledge graph for further investigation, the company would like to convert this information into RDF. Company B is using the GoodRelations Vocabulary to describe its products. For example, a fraction of the sample product page is given as follows.

Listing 2.3 A fraction of the sample product page

```
<div xmlns:rdf="http://www.w3.org/1999/02/22-rdf-syntax-ns
    #"
     xmlns="http://www.w3.org/1999/xhtml"
     xmlns:foaf="http://xmlns.com/foaf/0.1/"
     xmlns:gr="http://purl.org/goodrelations/v1#"
     xmlns:xsd="http://www.w3.org/2001/XMLSchema#">

  <div about="#offering" typeof="gr:Offering">
    <div property="gr:name" content="CompanyB Laptop Model
        123" xml:lang="en"></div>
    <div property="gr:description" content="A classic,
        timeless design
Providing a smooth computing experience
Precise, crystal clear audio to complete the cinematic
    experience
Access your data and files anytime, anywhere" xml:lang="en
    "></div>
    <div rel="foaf:depiction"
         resource="http://companyB.com/previews/LaptopM123/
             images/intro.jpg">
    </div>
    <div rel="gr:hasBusinessFunction"resource="http://purl
        .org/goodrelations/v1#Sell">
    </div>
    <div rel="gr:hasPriceSpecification">
      <div typeof="gr:UnitPriceSpecification">
        <div property="gr:hasCurrency"content="GBP"
            datatype="xsd:string"></div>
        <div property="gr:hasCurrencyValue"content="599"
            datatype="xsd:float"></div>
      </div>
    </div>
    <div rel="gr:acceptedPaymentMethods"
         resource="http://purl.org/goodrelations/v1#PayPal
             "></div>
    <div rel="gr:acceptedPaymentMethods"
         resource="http://purl.org/goodrelations/v1#
             MasterCard"></div>
    <div rel="foaf:page"resource="http://companyB.com/page
        /laptopM123/"></div>
  </div>
</div>
```

The product metadata is specified using the GoodRelations Vocabulary in an RDFa syntax.[22] To use the GRDDL standard for extracting the laptop information in RDF, Company A will need to inject two pieces of information into the crawled Web page. The first piece of information needed is to declare that the page contains GRDDL metadata. For an XHTML document, GRDDL reuses the *profile* construct of an HTML page. As shown in Line 2 of the following HTML code, by setting the *profile* attribute of the *head* element to the value of http://www.w3.org/2003/g/data-view, this page is declaring that (part of) its metadata can be extracted into RDF using GRDDL.

```
<!DOCTYPE html PUBLIC "-//W3C//DTD XHTML 1.1//EN""http://
    www.w3.org/TR/xhtml11/DTD/xhtml11.dtd"><html xmlns="
    http://www.w3.org/1999/xhtml" xml:lang="en"lang="en">
  <head profile="http://www.w3.org/2003/g/data-view">
    <title>Company B's Laptop Model 123 Product</title>
  </head>
  <body>
  ...
```

After mentioning that the page is GRDDL compatible, the next point would be to specify what the transformation algorithm is. In our example, we use XSLT, the XML transformation language,[23] which is the most common approach to transform an XML document into another format. Line 3 of the following HTML code illustrates how to specify the XSLT template in GRDDL for transforming the metadata. Given that the product metadata in our example is specified using an RDFa format, in this example, we are using a general XSLT template for converting an RDFa into an RDF format: http://www.w3.org/2008/07/rdfa-xslt. In practical cases, the users should choose suitable transformations or define new ones according to the syntaxes of metadata to be converted.

```
<!DOCTYPE html PUBLIC "-//W3C//DTD XHTML 1.1//EN""http://
    www.w3.org/TR/xhtml11/DTD/xhtml11.dtd"><html xmlns="
    http://www.w3.org/1999/xhtml" xml:lang="en" lang="en">
  <head profile="http://www.w3.org/2003/g/data-view">
    <title>Company B's Laptop Model 123 Product</title>
    <link rel="transformation" href="http://www.w3.org
        /2008/07/rdfa-xslt"/>
  </head>
  <body>
  ...s
```

After explicitly specifying the transformation algorithm, we have successfully updated the GRDDL setting to be compliant with our product example. The result HTML page is available at http://www.kdrive-project.eu/kgbook/kgboo-grddl.html.

[22] http://www.w3.org/TR/xhtml-rdfa-primer/.

[23] http://www.w3.org/TR/xslt.

To execute the transformation, the only point left is to find a software or a service which supports the GRDDL extraction. There are quite some tools or libraries out there. This page, http://www.w3.org/wiki/GrddlImple-mentations, lists some of the popular implementations of the GRDDL standard. For conducting a test run of our example, one can use the online GRDDL service of `librdf.org` by providing it with our result HTML page.[24] There is a shortcut you can directly open in your browser to see the output of our example: http://librdf.org/parse?language=grddl&uri=http%3A%2F%2Fwww.kdrive-project.eu%2Fkgbook%2Fkgboo-grddl.html&content=&Run+Parser=Run+Parser&.cgifields=language. If everything goes well, you will see a result page, which lists a table with all the triples converted. The top several rows should look exactly the same as Fig. 2.13.

Subject	Predicate	Object
http://www.kdrive-project.eu/kgbook/kgboo-grddl.html#offering	http://xmlns.com/foaf/0.1/page	http://companyB.com/page/laptopM123/
http://www.kdrive-project.eu/kgbook/kgboo-grddl.html#offering	http://purl.org/goodrelations/v1#acceptedPaymentMethods	http://purl.org/goodrelations/v1#MasterCard
http://www.kdrive-project.eu/kgbook/kgboo-grddl.html#offering	http://purl.org/goodrelations/v1#acceptedPaymentMethods	http://purl.org/goodrelations/v1#PayPal
blank node r1422458441r2615r1	http://purl.org/goodrelations/v1#hasCurrencyValue	599^^<http://www.w3.org/2001/XMLSchema#float>
blank node r1422458441r2615r1	http://purl.org/goodrelations/v1#hasCurrency	GBP^^<http://www.w3.org/2001/XMLSchema#string>
blank node r1422458441r2615r1	http://www.w3.org/1999/02/22-rdf-syntax-ns#type	http://purl.org/goodrelations/v1#UnitPriceSpecification

Fig. 2.13 The snapshot of a GRDDL example result

2.4 Knowledge Graph Versus Linked Data

In many literature, RDF datasets, Linked Data and Knowledge Graph are often mentioned in the same contexts. It is important to clarify this terminology and to highlight the major differences between the three related concepts.

When we talk about RDF datasets, we mean data collections where the data is presented in an RDF format, i.e. in relational structures consisting of a *subject*, a *predicate* and an *object*, where the former two are identifiers (URIs) and the latter is either an identifier or a value. With Linked Data, we refer to multiple RDF datasets, developed, maintained and distributed independently of one another, and yet inter-linked. Cross-links may result from referencing the same individuals with the same URIs in different datasets, or may be supplemented by drawing mappings, e.g. using the *owl:sameAs* predicate.

A knowledge graph is a structured dataset that is compatible with the RDF data model and has an (OWL) ontology as its schema. A knowledge graph is not necessarily linked to external knowledge graphs; however, entities in the knowledge graph usually have type information, defined in its ontology, which is useful for providing contextual information about such entities. Knowledge graphs are expected to

[24]http://www.kdrive-project.eu/kgbook/kgboo-grddl.html.

Table 2.3 RDF Datasets versus Linked Data versus Knowledge Graph

Features	Pure RDF datasets	Linked Data	Knowledge Graph
Machine readability	Y	Y	Y
Human readability	NN	NN	Y
Data distribution	N	Y	NN
Inter-dataset linkage	L	Y	Y
Data integration	NN	NN	Y
Data consistency	NN	NN	Y
Reliability	NN	NN	Y
High quality	NN	NN	Y

be reliable, of high quality, of high accessibility and providing end user oriented information services.

Table 2.3 compares the features of pure RDF Datasets (without using OWL), Linked Data and Knowledge Graph within several dimensions, where *Y* means *Yes*, *L* means *Limited*, *N* means *No* and *NN* means *Not Necessarily*. For example, pure RDF Datasets allow only limited inter-dataset linkages, as RDF does not support sameAs; however, Linked Data and Knowledge Graph can have inter-dataset linkages, thanks to *owl:sameAs*. Linked Data is not necessarily good for data integration because many Linked Data might not have schema/ontology defined.

2.5 Knowledge Graph for Web Searching and Knowledge Graph for Enterprise

As a brand, the *Google Knowledge Graph* has been used as a knowledge base to enhance the Google search engine results with semantic search information and become popular with Google by May 2012. It is natural for the reader to think of the Google's Knowledge Graph and wonder what the relation is between Google's with ours in this book. In this part, we review the Knowledge Graph for Web Searching and the Knowledge Graph for Enterprise, and then compare them in a brief and high level.

Knowledge Graph for Web Searching

The Google Knowledge Graph provides a short summary about the topic with structured information, as well as a list of commonly used links to the other sites in order to back up for the most possible queries on that topic.

It is based on two main technologies: an ontology with a pretty noncomplicated hierarchy of types, plus a process for data gathering and integration. In the former, schema.org[25] provides a collection of shared vocabularies in a hierarchy of about 700 types,[26] which are to be used by Web-builders to mark up their pages so that the pages can be understood and indexed by the major search engine Google, as well as Microsoft Bing, Yandex and Yahoo!

The latter one, data gathering, involves Web-scale distributed data annotation activities, based on some open encyclopaedic sources and on the provided APIs. Dated by June 2015, encyclopaedia used by the Google Knowledge Graph includes: the CIA World Factbook,[27] which is an almanac-style reference resource about the countries of the world; Wikipedia,[28] which is a well-known free-access and free-content Internet encyclopaedia; Freebase,[29] which is an online collection of structured data about common information (e.g. well-known people, places and things), and provides APIs for users to annotate their Web pages; and Wikidata,[30] which provide a common source of certain data types (e.g. birth dates) which can be used by Wikimedia sister projects such as Wikipedia, Wikivoyage, Wikisource and others.

Knowledge Graph for the Other Search Engines

Knowledge graphs have also been used in some other search engines. In Microsoft Bing, since 2013, the RDF-based Knowledge Base Satori has been helping Bing to easily identify queries related to well-known entities, including people, places and organisations. In Yahoo! a platform *Yahoo! Knowledge* provides services of a Yahoo-version knowledge graph of entities and concepts, in order to support knowledge-based applications at Yahoo! which includes Web Search, Media Verticals, Advertisement, etc.

Knowledge Graphs for Enterprise Information Services

Managing many kinds of enterprise knowledge is one of the most relevant businesses for the IT industry. This is generally achieved by creating data infrastructures, developing applications working on them, managing processes for data acquisition, curation, maintenance, integration and access. Relational database management sys-

[25]Full Hierarchy of the schema is found at: http://schema.org/docs/full.html.

[26]The number of types mentioned here dates back to June 2015.

[27]The World Factbook is produced by the United States Central Intelligence Agency for the US policymakers. https://www.cia.gov/library/publications/the-world-factbook/.

[28]http://wikipedia.org.

[29]http://www.freebase.com/. Freebase was merged into Wikidata by 30 June 2015, and Freebase APIs were replaced by Wikidata APIs.

[30]http://wikidata.org.

tems, along with data integration and warehousing platforms, are at the basis of such an industry.

Knowledge graphs provide new paradigms and platforms for implementing a variety of enterprise knowledge bases in a more effective and powerful way. For enterprise organisations, implementing knowledge graphs means implementing their architecture and supporting their acquisition and maintenance processes. Organisational boundaries make it possible (albeit nontrivial in many cases) to effectively support key prerequisites of such processes.

Primarily, organisations may provide unified knowledge schemas (ontologies) for their business and spread them along corporate branches. These ontologies can be built on the basis of existing business vocabularies and industry models, and are easily workable to follow the business evolution. Generally backed by RDF and property graph stores (or simply graph data stores) [11], knowledge graphs provide better schema evolution capabilities with respect to relational platforms (cf. Sect. 3.3). This facilitates the adoption of shared enterprise conceptual models, thus preventing the proliferation of heterogeneous schemas for the same business entities, which is one of the major causes of inefficiency of information management within a large business organisation.

As regards singular business entities and consumers on the Web, enterprises can leverage structured organisational processes and work with controlled collections of information sources, like in data warehouse systems. As in traditional enterprise information management processes, knowledge graphs require the consolidation of datasets which are produced independently in different branches and business units, in a centralised repository for subsequent consumption. However, encoding datasets with a fixed layout (e.g. RDF triples), and based on a single shared schema, greatly facilitates integration and data quality assurance.

The problem of creating centralised knowledge out of distributed and independent sources is one of the major challenges in the Semantic Web endeavour. Even if large organisations exhibit variety in the way concepts and facts are interpreted in different contexts, tackling such *epistemic* problems appears to be easier at the enterprise level.

In summary, we realise that the knowledge, or schema or ontology, embedded in the knowledge graphs for Web searching is more generic and simpler common knowledge, but the one for enterprise is more domain-specific. Furthermore, we note that for a Web-scale searching task, it is obviously very difficult to apply a more expressive knowledge representation language, owing to the prohibitive costs in data maintenance and computation. But this is not a big problem in the applications of a knowledge graph on an enterprise scale. Also considering the data and knowledge acquisition, it is harder and more difficult to control the quality of Web-scale annotation activities, but it is easier to gain enterprise data which is also more reliable.

Table 2.4 Knowledge Graph for Web search versus for enterprise

Features	KG for Web searching	KG for enterprise
Data source	Distributed	usu. centralised
Openness of data	Open to public	Private
Size of data	Huge	Big
Data acquisition	Harder	Easier
Quality of data	Low	High
Ontology language	Simple	Likely to be more expressive
Knowledge	Generic knowledge	Domain-specific knowledge

It seems that Google, Yahoo! and Bing are using simpler applications of Knowledge Graphs, for search engines, they are good enough. Applications of Knowledge Graph in an enterprise, on the other hand, can provide more services than just "searching," as such applications are able to organise higher quality data and more expressive and meaningful knowledge. In Table 2.4, we compare the knowledge graph for enterprise and knowledge graph for Web searching.

Chapter 3
Knowledge Architecture for Organisations

Ronald Denaux, Yuan Ren, Boris Villazon-Terrazas, Panos Alexopoulos, Alessandro Faraotti and Honghan Wu

In this chapter, we prepare you a high-level overview of what is needed in order to create, maintain and exploit knowledge graphs for a real application. We realise that there is no one way of doing this for all organisations and all use cases of knowledge graphs; hence in this chapter, we introduce an **Abstract Reference Architecture** (ARA) that will help you to understand the main phases and tasks required during the lifecycle of knowledge graphs. For each of the phases and tasks in this ARA, we then present a more detailed description of possible approaches, methodologies and tools which have been reported in the literature. By the end of this chapter you should have a good idea of the tasks you would likely encounter when building and maintaining knowledge graphs. You will also have a better understanding of how knowledge graphs can be used within large organisations.

R. Denaux (✉) · B. Villazon-Terrazas · P. Alexopoulos
Expert System, Prof. Waksman 10, 28036 Madrid, Spain
e-mail: rdenaux@expertsystem.com

B. Villazon-Terrazas
e-mail: bvillazon@expertsystem.com

P. Alexopoulos
e-mail: palexopoulos@expertsystem.com

Y. Ren
University of Aberdeen & King's College, Aberdeen AB24 3UE, UK
e-mail: y.ren@abdn.ac.uk

A. Faraotti
IBM Italia, via Sciangai 53, 00144 Rome, Italy
e-mail: alessandro.faraotti@it.ibm.com

H. Wu
King's College London, De Crespigny Park, London SE5 8AF, UK
e-mail: honghan.wu@kcl.ac.uk

J.Z. Pan et al. (eds.), *Exploiting Linked Data and Knowledge Graphs in Large Organizations*, DOI 10.1007/978-3-319-45654-6_3

3.1 Architecture Overview

In very broad terms, there are three main tasks related to the use of knowledge graphs: construction, storage and consumption. These tasks are the main layers in our ARA depicted in Fig. 3.1.

Knowledge Acquisition and Integration Layer Firstly, you need to **create** the Knowledge Graph. This can be done bottom-up—starting from the data that you already have in order to extract the knowledge in terms of entities and their relations—top-down—analysing the various use cases you have and defining the knowledge you need to gather data about—or middle-out—essentially, combining the bottom-up and top-down approaches. This task corresponds to the *Knowledge Acquisition and Integration layer* in the reference architecture (Chaps. 4 and 5 of this book).

Knowledge Storage layer Secondly, you need to **store** the knowledge graph in such a way that you will be able to evolve the graph as time goes by (e.g. by adding new types of knowledge) and access the knowledge encoded in the graph in an efficient manner. This task corresponds to the *Knowledge Storage layer* in the architecture.

Knowledge Consumption layer Finally, you need to put the knowledge encoded in your knowledge graph to **use within your organisation** in order to improve the efficiency of your organisation. This task corresponds to the *Knowledge Consumption layer* in the architecture (Chaps. 6 and 7 of this book).

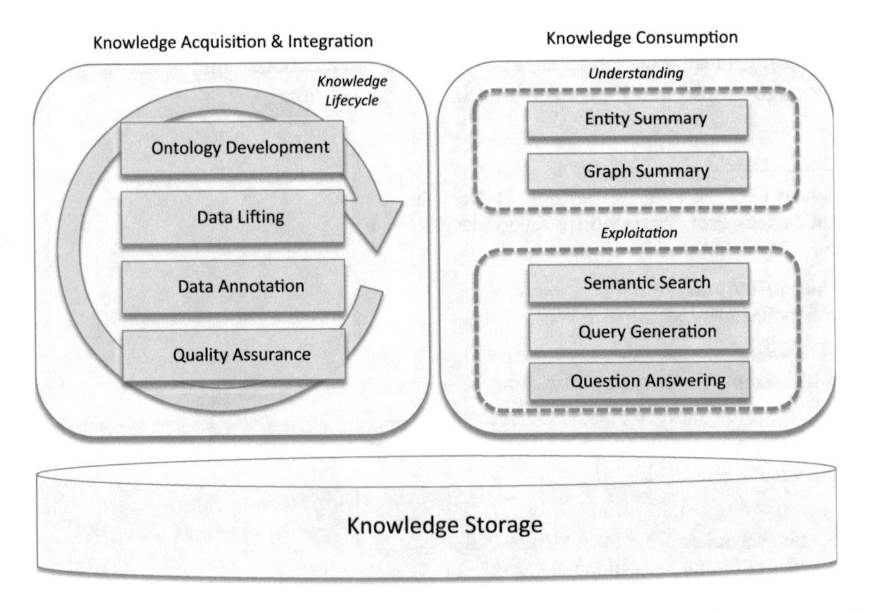

Fig. 3.1 The **Abstract Reference Architecture** (ARA) for Knowledge Graph management in organisations

Each of the layers in the abstract architecture can consist of many subtasks, which can be performed in different ways and using different technologies. The remaining chapters in this book will focus on specific subtasks in great detail; hence in this chapter, we will focus on introducing the general approaches and giving some short examples of how these tasks can be and have been performed.

In Sect. 3.2, we discuss the various subtasks of knowledge acquisition and integration. We will start by pointing to various methodologies for managing the lifecycle of knowledge graphs, i.e. for deciding when you will perform certain subtasks and at what level of completion: from analysing the data (when using a bottom-up approach) or your use-case domain (top-down), through modelling the domain data and generating the data, up to publishing and exploiting the knowledge graph. As we will see, there are various alternatives on how to do this; which one is the best for you will depend on your organisation: which data you have at your disposal, how interconnected that data already is and what your intended usage is.

In Sect. 3.2.1 we will also introduce various methodologies for modelling the schema of your knowledge graph in detail, i.e. for formally defining all of the entity types that you will include (or that are already present) in your knowledge graph. Having such formal definitions of your domain can be useful for automatically interpreting the data in your knowledge graph. As we will see later, having a rich description of your domain can be used by various algorithms to detect inconsistencies or to perform disambiguation during textual annotation. Formal definitions can also be detrimental, because they can overly restrict the data: e.g. by detecting inconsistencies when it is not the data which is wrong, but rather the schema which is not flexible enough to consider some valid use cases.

One of the main subtasks of a knowledge graph creation is the re-use of the existing structured data and schemas in your organisation. In Sect. 3.2.2 we introduce various approaches for converting the existing resources such as relational databases (and their schemas), thesauri or structured documents.

For unstructured documents (e.g. office documents, emails), the approach is different, because you often need to use Natural Language Processing in order to be able to link such documents to your knowledge graph. In Sect. 3.2.3 we introduce how this can be done and point to how various systems perform this task.

To wrap up our discussion on knowledge acquisition, we will also mention how parts of the domain definitions can be learned from the existing data you may have. This is called *Ontology Learning* and is discussed in Sect. 3.2.4.

Regarding the subtasks required for knowledge storage, the standard way to store knowledge graphs is, unsurprisingly, using graph databases. However, we will first discuss how to avoid duplicating the storage of data you already store in some other format (e.g. as relational databases) by adding a translation layer in order to access those datasources using the conceptual layer defined by your knowledge graph. This means that you can ask your knowledge graph for entities of a type X and these entities can be retrieved from an existing relational database. The disadvantage of this approach is that some of the automatic reasoning that can be performed on graph databases cannot be (efficiently) performed by the translation layer. More on this approach will be discussed in Sect. 3.3.1.

As we said, the typical way of storing knowledge graphs is by using graph databases. In Sect. 3.3.2 we will introduce a standardised version of graph databases built on top of Web and Linked Data technologies called RDF Stores. These databases were developed intensively during the last decade, and as such we can draw on an extensive number of reported experiences in the literature (and our own) regarding systems built using such databases. We introduce various RDF stores which are available and discuss how such RDF stores can be evaluated across various dimensions. Finally, in Sect. 3.3.3 we will also analyse a different approach for graph databases, which is not standardised and discuss how it compares to RDF stores.

For the final layer of architecture, the consumption of knowledge from knowledge graphs, in Sect. 3.4 we will introduce various typical uses of knowledge graphs: from semantic search (i.e. improving your current search interfaces), through provision of context to your employees or customers via entity and graph summaries, to answering questions relevant to your organisation.

3.2 Acquisition and Integration Layer

3.2.1 Ontology Development

One of the main steps during the creation of a knowledge graph is to decide what knowledge you want to capture in your graph. This is part of the data modelling activity. At this stage you can decide to use a lightweight existing vocabulary, but you can also decide to use a more heavyweight schema.[1] With lightweight schemas you only name the main relations between the entities in your data (you don't even need to name the types of entities). A more heavyweight schema will contain formal definitions with restrictions on what specific entity types mean in your context. Heavyweight schemas can be used by some algorithms to (partially or fully) automate some tasks later on (such as data interlinking or text annotation and disambiguation). If you decide to create formal definitions for your entities, you can reuse several decades of experience in *ontology development* (the technical term for creating the formal definition of a domain).

The importance of applying engineering principles to the ontology development process has been recognised for a long time now and several methodologies have been proposed for this purpose including METHONTOLOGY [75], Diligent [248], HCOME [138], NeOn [226] and DOGMA [150]. Typically, such a methodology defines a set of activities that need to be performed while developing an ontology and usually suggests or provides methods and techniques for effectively carrying out these activities' tasks.

Moreover, the practice of developing dedicated tools to cope with specialised aspects and dimensions of the ontology development process has been exemplified

[1] Note that you can always decide to start with a lightweight vocabulary and add formal definitions as needed (even after you have been using your knowledge graph successfully for initial use cases).

by several relevant works. For example, OntOWLClean [251] is an OWL-based tool that facilitates the easier and more intuitive application of OntoClean [95], a well-known methodology for the evaluation of the ontological adequacy and logical consistency of taxonomic relationships. Another related tool is OntoParts [128] that helps ontology authors decide which particular types of part–whole relations (material–object, portion–object, place–area, etc.) are appropriate for their ontology. An evolution of OntoParts is FORZA (Foundational Ontology and Reasoner-enhanced axiomatiZAtion) [129] that also supports the task of linking a domain ontology to categories of the DOLCE (Descriptive Ontology for Linguistic and Cognitive Engineering) foundational ontology [155].

Authorisationwise, ontologies are either authorised (semi-)manually by domain experts or knowledge engineers, for example, Sect. 4.2 of Chap. 4 presented a competency question-driven approach that facilitates the authorisation process; or automatically generated from the data or knowledge repositories, e.g. Sect. 5.2 of Chap. 5 presents a schema learning algorithm based on the Bayesian Description Logic Network.

3.2.2 *Ontologisation of Non-Ontological Resources*

You can define the schema for your knowledge graph from scratch by analysing your use cases or your domain; this is the top-down approach. However, you can also try to reuse schemas about existing databases or classification schemas in your organisation (or existing public schemas). The main problem here is that you need to translate these schemas into RDF or OWL. This process can typically be mostly automatised because the existing schemas and classifications are *lightweight*, and thus can be captured faithfully by RDF or OWL.[2]

During the last decade, specific methods, techniques and tools were proposed for building ontologies from existing knowledge resources. Non-ontological resources (NORs) [246] are those knowledge resources whose semantics have not yet been formalised explicitly by means of ontologies. Examples of NORs are classification schemes, thesauri, lexical and folksonomies, among others. This type of resource encodes different types of knowledge and can be implemented in different ways. According to Hepp et al. [112, 113, 115] employing methods and techniques when ontologilising NORs to the level of ontologies is the key for the success of semantic technology for two main reasons: (1) if the use of semantic technologies for real-world data integration challenges is required, it is possible to refer to the original conceptual elements, and (2) for many domains, the existing category systems, XML schemas and normative entity identifiers are the most efficient resources for engineering ontologies.

The ontologilisation of NORs has led to the design of several specific methods, techniques and tools [85, 87, 115, 123]. These include methods for building ontologies from (1) **classification schemes** [101, 115], (2) **folksonomies** [5, 152], (3) **lexica**

[2]There is some initial work on extracting some heavyweight descriptions of the domain from the existing data, as we will see in Sect. 3.2.4.

[85, 86, 238], (4) **thesauri** [99, 100, 123, 139, 218, 239, 240, 252], (5) **databases** [20, 21, 223], (6) **XML** [10, 57, 87] and (7) **flat files** [82].

3.2.3 Text Integration via Named Entity and Thematic Scope Resolution

Integrating texts into knowledge graphs is typically done by means of two tasks, namely **Named Entity Resolution** and **Thematic Scope Resolution**. The first task involves detecting mentions of named entities (e.g. people, organisations or locations) within texts and mapping them to their corresponding entities in the knowledge graph source. For example, in the text *"Siege of Tripolitsa took place in Tripoli with Theodoros Kolokotronis being the leader of the Greeks. This event marked an early victory for the fight for independence from Turkey but it was also a massacre against the Muslim and Jewish population of the city"* one may identify, among others, the DBpedia entities http://dbpedia.org/resource/Tripoli, http://dbpedia.org/resource/Greece and http://dbpedia.org/resource/Turkey. The typical problem in this task is ambiguity, i.e. the situation that arises when a term may refer to multiple different entities. For example, "Tripoli" may refer, among others, to the capital of Libya or to the city of Tripoli in Greece.

On the other hand, the **Thematic Scope** of a document can be defined as the set of semantic entities the document actually talks about. For example, the scope of a film review is typically the film the review is about while a biographical note's scope includes the person whose life is described. It is important to notice that resolving this scope is not the same as performing a named entity resolution, the reason being that the thematic scope of a document is not always equivalent to the semantic entities it contains. As an example of this, consider the following text from a film review: *"Annie Hall is a much better movie than Deconstructing Harry, mainly because Alvy Singer is such a wellformed character and Diane Keaton gives the performance of her life. I even think it is better than Manhattan*". The result of named entity resolution in this text (using a film ontology) would be the extraction of the entities *"Annie Hall"*, *"Deconstructing Harry"* and *"Manhattan"*. Yet, as one can easily infer from the text, the thematic focus of the text is not the film *"Deconstructing Harry"* but rather *"Annie Hall"*.

Section 5.1 (p. 115) gives the technical details of the Named Entity and Thematic Scope Resolution from unstructured data.

Named Entity Resolution Frameworks

An ontology-based entity disambiguation approach is described in [133] where an algorithm for entity reference resolution via spreading activation on RDF graphs is proposed. The algorithm takes as input a set of terms associated with one or more ontology elements and uses the ontology graph for spreading activation in order to compute Steiner graphs, namely graphs that contain at least one ontology element for each entity. These graphs are then ranked according to some quality measures and the highest ranking graph is expected to contain the elements that correctly correspond to the entities.

Another approach put forth is that of [94] where the application of restricted relationship graphs (RDF) and statistical NLP techniques to improve named entity annotation in challenging Informal English domains is explored. The applied restrictions are
(i) domain ones where various entities are a priori ruled out and (ii) real-world ones that can be identified using the metadata about entities as they appear in a particular post (e.g. that an artist has released only one album or has a career spanning more than two decades).

In [106] Hassel et al. propose an approach based on the DBLP ontology which disambiguates authors occurring in mails published in the DBLP-mailing list. They use ontology relations of length one or two, in particular the co-authorship and the areas of interest. Also, in [203] the authors take into account the semantic data's structure, which is based on the relations between the resources and, where available, the human-readable description of a resource. Based on these characteristics, they adapt and apply two text annotation algorithms: a structure-based one (Page Rank) and a content-based one.

Several approaches utilise Wikipedia as a highly structured knowledge source that combines annotated text information (articles) and semantic knowledge (through the DBpedia[3] ontology [14] and YAGO [227]). For example, DBpedia Spotlight [157] is a tool for automatically annotating mentions of DBpedia resources in text by using (i) a lexicon that associates multiple resources to an ambiguous label and which is constructed from the graph of labels, redirects and disambiguations that DBpedia ontology has and (ii) a set of textual references to DBpedia resources in the form of Wiki links. These references are used to gather textual contexts for the candidate entities from Wikipedia articles and use them as disambiguation evidence.

A similar approach that uses the YAGO is the AIDA (Advanced Image and Data Acquisition) system [117], which combines three entity disambiguation measures: the prior probability of an entity being mentioned, the similarity between the contexts of a mention and a candidate entity, and the semantic coherence among candidate entities for all mentions together. The latter is calculated based on the distance between two entities in terms of type and subclassOf edges as well as the number of incoming links that their Wikipedia articles share.

Thematic Scope Resolution Frameworks

Thematic scope resolution frameworks can be seen as tag recommendation systems based on domain ontologies. In the literature, there are many examples of tag recommender systems [88, 103, 161, 182, 229], but only few of them use ontologies.

[3] http://dbpedia.org.

A work in which ontologies are used for tagging is that of [230] where the authors present ePaper, a system that uses a hierarchical news ontology as a common language for content-based filtering in order to classify news items and to deliver personalised newspaper services on a mobile reading device. In another work [193], the authors propose a tag recommendation process based on key phrase extraction and ontology reasoning. In particular, their approach involves the utilisation of linguistic and statistical processing for determining key phrases that could be potential tags and the exploitation of domain ontologies for suggesting tags that are not present within the document. For the latter, they use a reasoning mechanism based on the subsumption relationship between concepts (is-a) and the spreading activation algorithm of [195].

A similar approach is presented in [174] where the authors discuss an ontology-based document annotation for the purpose of semantic indexing and retrieval. The method they propose expands, both syntactically and semantically, concept descriptions taken from the domain ontology in order to enhance matching in the retrieval process. The syntactic expansion is based on lexical resources (e.g. Wordnet) while the semantic one on a concept exploration algorithm that is applied to the ontology.

In [30] the authors propose GoNTogle, a framework for document annotation and retrieval, built on top of Semantic Web and Information Retrieval technologies. For the annotation part, GoNTogle supports the automatic annotation of a whole document or parts of it with ontology concepts through a learning method based on *weighted kNN* classification [163] that exploits user annotation history and textual information to automatically suggest annotations for new documents.

In [6] the authors suggest an approach to generate semantic tag recommendations for documents based on Semantic Web ontologies and Web 2.0 services. In particular, their proposed process starts with the extraction of document entities through the utilisation of Web 2.0 services (such as Yahoo's Term Extraction service and their transformation into a topic map using SKOS vocabulary (Simple Knowledge Organisation System) [159]. Then, the topics of this topic map are matched, based on document classification methods, to instances of some domain ontology expressed according to the PIMO (Personal Information Models) ontology [204]. The matching pairs are shown to the users as tag recommendations and they decide whether to accept or to reject them.

3.2.4 Ontology Learning

In cases when you do not have existing schemas available (e.g. databases or classification schemas), but you still want to avoid creating a heavyweight schema (i.e. an ontology) from scratch, you may want to look into the possibility of learning the ontology from (unstructured) data. This is still a topic being researched, but we give an overview of the current state of approaches and systems, since they are a promising direction for improving the creation of knowledge graphs.

Ontology learning [53, 261] seeks to discover ontological knowledge from various forms of data [146], either automatically or semi-automatically, in order to overcome the bottleneck of knowledge acquisition in ontology development. Many works on this topic try to tackle specific tasks, such as concept and relation extraction, extending existing ontologies and ontology population [26, 181, 228].

For example, in [181] a methodology for semi-automatic ontology extension by glossary terms based on text mining methods and considering ontology content, structure and co-occurrence information is proposed. Another automated ontology extension system is SOFIE (A Self-Organizing Framework for Information Extraction) [228] that is able to parse natural language documents, extract ontological facts from them and link the facts into an ontology. Other works focus on learning more complex knowledge such as concept hierarchies. For example, in [55] this is done by means of the formal concept analysis while in [249, 257] by means of the latent Dirichlet allocation learning algorithm.

3.3 Knowledge Storing and Accessing Layer

Once you have identified the data relevant to your organisation that you want to include in your knowledge graph, the next step is to find out how to access or store that data, which is the topic of this section. The goal is to provide access to the data to the various tools which can be used to exploit the knowledge that has been captured in the graph. In this section, we discuss the two main architectural options.

- in Sect. 3.3.1 we present how to simply reuse whatever storage your organisation currently uses by defining bridges in order to access the data as if it were stored in a graph database.
- in Sects. 3.3.2 and 3.3.3 we discuss two approaches for storing graph data. The first one is based on the RDF data model standard which was introduced in Sect. 2.1.1, while the second one uses an alternative data model which has not been standardised.

3.3.1 Ontology-Based Data Access

Large organisations are producing an ever-growing amount of evolving information which is often contained within 'silos' of data, stored in different systems (DB, applications, CMS...) and represented in heterogeneous formats (documents, media, tables). Since the 1990s the need to unlock and integrate such information was identified and has become known as the *Data Integration problem*. Traditional data integration systems developed over the last two decades have been designed to extract, clean and reconcile data into a unique place, either a master data repository or a warehouse system. Although very efficient and largely employed, those systems do not explicitly deal with the semantics of the managed information and usually require to materialise a new copy of the data. This means that the resulting data warehouse is difficult to interpret and use, and it requires additional storage solutions.

Ontology-Based Data Access (OBDA) is a novel paradigm in which the access to distributed, heterogeneous and typically incomplete data sources is mediated by domain-oriented conceptual models expressed through formal ontologies. In an OBDA system, the ontology acts as a semantic layer that combines and enriches the information stored in data sources into a unified view, which represents not only how data are organised, but also what is their intended meaning. OBDA supports reasoning and query answering over conceptual models, where users and applications can transparently query the ontologies without having to cope with underlying schemes and data structures. This approach leverages on logic languages specifically tailored to allow a clear separation between the semantic level (terminological level or TBox) and data level (assertional level or ABox). OBDA systems perform reasoning only against the TBox without relying on data assertions, consequently those systems are scalable with respect to the size of data because the data itself does not affect reasoning. Moreover, such separation allows to virtualise the ABox. Real data can be maintained into sources that are externally and independently managed as part of legacy applications. OBDA is therefore tailored to address the Semantic Data Integration problem [144] of integrating data at conceptual level through a virtual reconciled view.

In Fig. 3.2 a generic architecture of an *OBDA System* providing both users and other applications with semantic services to reason and to perform queries against the ontology and data sources is depicted. The system is configured with one or more ontology documents, representing the conceptual model TBox of the domain, together with a set of declarative mappings that establish relations between the conceptual model and data sources' schemes. A generic OBDA System can be logically represented through three macro components: TBox, ABox and Semantic Services. The TBox includes an *Ontology Manager* component which stores the conceptual model and optionally keeps such model up to date; if the input ontology contains unsupported axioms the Ontology Manager may leverage on an *Approximator* to reduce the ontology expressivity. The ABox manages the ontology extensional level leveraging on the mappings to access external data sources. It can be implemented by following either a virtual or a materialised approach. In the *Virtual ABox* approach, the *Data Access* component uses the mappings to retrieve ontology instances directly from the data sources upon each request. In the *Materialised ABox* approach, the ontology instances are kept with the system in a local repository (in a database or in a triple store), the *Materialiser* component leverages on the mappings to extract and periodically update the ontology instances. *Semantic Services* implements system logic and includes a *Reasoner* that implements basic reasoning tasks such as subsumption computation leveraging on the TBox and a *Query Reformulator* component that reformulates (conjunctive) queries leveraging on the Reasoner and forwards them to the ABox which is in charge of their execution.

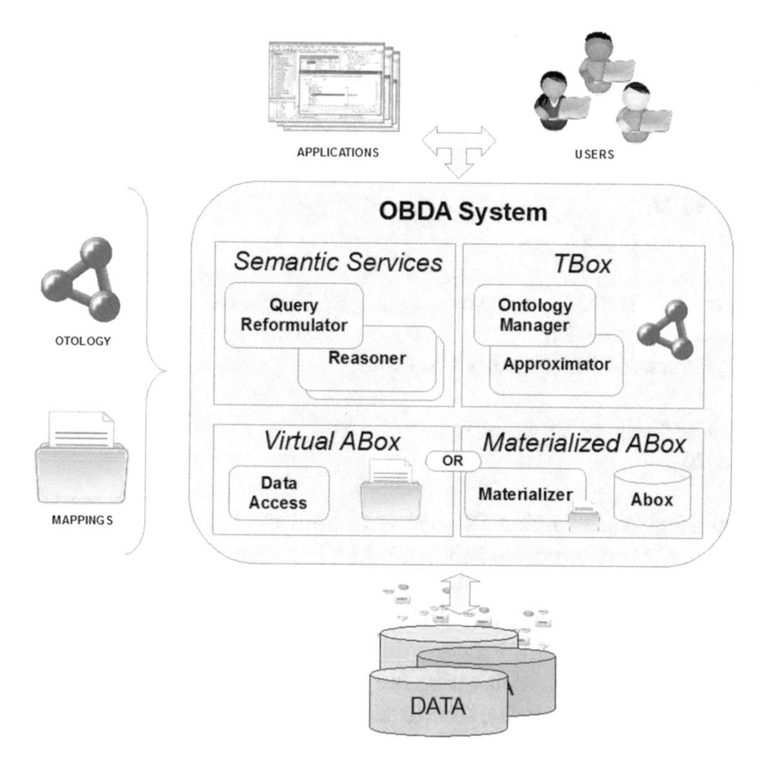

Fig. 3.2 High-level architecture

Ontologies and Reasoning

Ontologies are usually based on a logical formalism called *Description Logics* (DLs) [17], which are decidable fragments (i.e. guaranteed to be computable) of the First-Order Logic characterised by well-understood computational properties. DLs are commonly used to describe conceptual models that are used as *global views* in semantic data integration systems [144]. Characteristics that have been proven for such Description Logic languages allow us to identify theoretical properties of the system as a whole. Below, we discuss in technical detail why OBDA systems are expected to have a similar performance as traditional relational databases, while at the same time being scalable and having sufficient expressivity to capture most current modelling languages such as Entity Relational model (used in traditional relational databases) and UML (Unified Modelling Language) (used in software development and specification).

In order to define ontologies, OBDA systems adopt DLs that derive from the DL-Lite family [13]. Those languages optimise the trade-off between expressiveness and reasoning algorithms' complexity in order to perform conjunctive query answering and subsequently other reasoning tasks reducible to it, while keeping the complexity of algorithms within LOGSPACE respect the size of the data; in other words, reasoning in DL-Lite has the same complexity of query answering on traditional relational databases. In the following, the DL-Lite F language described in [44] is reported:

Concepts:
$B ::= A \mid \exists R \mid \exists R^{-}$
$C ::= B \mid \neg B \mid C_1 \sqcap C_2$

TBox assertions are of the form
$B \sqsubseteq C$ *inclusion assertion*
$(functR)$, $(functR^{-})$ functionality assertions

ABox assertions are of the form
$B(a), R(a, b)$

The intuition underpinning DL-Lite is to limit the language expressivity so as to rely only on TBox axioms when reasoning. For example in [44] it is shown that supporting inclusion axioms like $\neg A \sqsubseteq G$ or $\exists R.C \sqsubseteq G$ increase the complexity of conjunctive query answering to coNP-hard.

Due to such a limitation, it turns out possible to separate the TBox, in which concepts, relationships and constraints are defined, from the ABox containing individuals and assertions (the data), so as to achieve scalability and ABox virtualisation; although limited DL-Lite is powerful enough to capture the most common modelling language's expressiveness (e.g. ER, UML).

DL-Lite is the basis of the OWL2-QL profile[4] described in Sect. 2.1.2.

Mapping and Data Sources

As OBDA systems aim to keep the data in their original databases (relational databases, document repositories, NoSQL stores, etc.), the mappings between the schema (i.e. the conceptual model) in the knowledge graph and the various schemas in the various databases play a central role. Below, we discuss how such mappings can be defined and what their impact is on the performance of OBDA systems.

The mapping M describes how the ontology, which represents the global schema G, and data sources' schema S are connected. M consists in a collection of assertions in the form $q_G \longleftarrow q_S$ each one composed by two queries, a query q_G over G and a query q_S over S.

In data integration two kinds of mappings can be distinguished depending on which schema is considered to be the central one. When the global schema is central, each single predicate of G is associated to a view on S and therefore q_G is an atomic query built on a single predicate of G, this kind of mapping is named GAV (Global as View). In the other case, when S is central, the mapping is named LAV (Local

[4]http://www.w3.org/TR/owl2-profiles/#OWL_2_QL.

as View) and q_S is an atomic query built on a single predicate of S, also called the local schema, and related to a query on the ontological schema G. It is also possible to have a hybrid kind of mapping named GLAV.

OBDA systems usually adopt the GAV approach because the conceptual model plays the central role and because query rewriting in GAV systems, also known as Unfolding, becomes a straightforward operation.

3.3.2 RDF Stores

As it is similar to natural language sentences, a triple-based data model is very simple and straightforward. Due to its superflexibility in portability and data schema, the triple-based data model has gained extreme popularity recently. More importantly, triples are adopted as the data model of the knowledge representations on the Web. The ontological data of RDF(S) and OWL is based on triples. Hence, triple stores (the data management systems for triple-based data) are the fundamental components for realising Linked Data based knowledge graphs.

Similar to other data management systems (DBMS), a triple store allows users to *store and query* triples. The main *performance indicators* of triple stores are focusing on the data volume, bulk loading speed and query answering efficiencies. Other aspects of DBMS such as transactions are also becoming prevalent in most triple stores. As knowledge base systems, the most important add-on of triple stores is the support of inference computations, which are implemented based on various formal semantics.

Below, we will first introduce a list of representative triple stores. Then, we will briefly discuss the performance and benchmarks. Finally, we will compare triple stores to other related NoSQL techniques and we will discuss a more recent, lightweight technique for storing and querying triples called Linked Data fragments.

Triple Store Systems

The popularity of triple-based data models is also reflected in the large number of implementations of triple stores, whose implementers range from Semantic Web communities to database practitioners, from academia to industry, from start-up companies to big players in the industry. In Table 3.1, we list some of these implementations which have been selected because they are up to date (having updates within the last two years) and comparable (having published benchmark results from implementers). A more comprehensive list can be found here.[5]

As shown in Table 3.1, triple stores can be implemented based on various techniques as follows.

- Relational Database
- Native Triple Store
- Graph Database

[5] http://en.wikipedia.org/wiki/List_of_subject-predicate-object_databases.

Table 3.1 List of selected triple stores

Triple stores	Implementation	Performance	APIs	License
Oracle spatial and graph	Object-relational	BM: LUBM4400[a]	Most languages	Commercial
		Max: 1.08 T		
		Load: 1.420 M/S		
		Query: 1.130 M/S		
		Infer: 1.527 M/S		
AllegroGraph	Graph database	BM: LUBM-like[b]	Most languages	Free edition commercial
		Max: 1.009 T		
		Load: 0.829 M/S Infer: RDFS		
Stardog	Triple store	BM: Unknown[c]	Java, Groovy	Community developer commercial
		Max: 50B		
		Load: 0.3 M/S		
		Infer: OWL2		
OpenLink Virtuoso v6.1	Relational property graphs	BM: Unknown[d]	Most languages	Open-source commercial
		Max: 15.4B+		
		Load: 0.275 M/S		
		Infer: *rdfs:subClassOf*, *rdfs:subPropertyOf*, *owl:sameAs*		
GraphDB 6.0	Graph database	BM: UNIPROT[e]	Java	Commercial
		Max: 12B		
		Load: 0.012 M/S		
		Infer: *pD**		
Jena TDB	Triple store	BM: UNIPROT[f]	Java	Apache license
		Max: 1.2B		
		Load: 0.225 M/S		
		Infer: RDFS, OWL-Lite		

[a] http://download.oracle.com/otndocs/tech/semantic_web/pdf/OracleSpatialGraph_RDFgraph_1_trillion_Benchmark.pdf
[b] http://franz.com/agraph/allegrograph/agraph_benchmarks.lhtml
[c] http://weblog.clarkparsia.com/2014/01/10/scalability-improvements-in-stardog-21/
[d] http://lod.openlinksw.com/
[e] http://www.ontotext.com/graphdb-benchmark-results/
[f] https://www.w3.org/wiki/LargeTripleStores

- Property Graph
- NoSQL
- Entity-centric

In the table, we also list the license information and supported programming languages of their APIs, both of which are important factors for choosing a triple store for knowledge graph construction.

Benchmarks and Performance

As we mentioned in the beginning of this subsection, the performance indicators of triple stores are mainly about the maximum volume of data, the bulk loading efficiency and the query answering efficiency. The *Performance* column of Table 3.1 lists the performance results of the triple stores. Before we explain the numbers and descriptions of the column, a brief introduction about benchmarks for triple stores is necessary.

Instead of targeting general triple stores, the benchmarks which we are aware of are mainly for evaluating RDF triple stores, which are mentioned in the following list:

- *LUBM and its extensions*: Lehigh University Benchmark (LUBM) [97][6] is almost the earliest and most widely used benchmark for Semantic Web repositories. LUBM provides an ontology describing universities (e.g. departments, staffs, students, etc.), a set of predefined queries and a data generator for creating potentially large test datasets. UOBM (The University Ontology Benchmark) [151] is an extension of LUBM focusing on inference and scalability, while Dave Kolas [135] extended LUBM for Spatial Semantic Web System.
- *General purpose*: Apart from LUBM, there are a number of other benchmarks designed for comparing RDF stores, RDF-mapped relational databases and other systems exposing SPARQL endpoints. Most of them are designed with specific domain or use case in mind. The Berlin SPARQL Benchmark (BSBM) [34] is a benchmark designed along an e-commerce use case. SIB (Social Intelligence Benchmark)[7] is designed for a social network domain. Linkbench [12] is based on the Facebook social graph.
- *Task-focused*: The third category is task-focused. These benchmarks are defined for specific tasks or aspects of an RDF store. FedBench [207] is designed for evaluating the federated SPARQL query processing. LODIB (Linked Open Data Integration Benchmark) [202] is designed for data translation and integration. JustBench [19] is designed for OWL reasoning tasks.
- *Real Queries and Data*: As used extensively by the community, DBpedia is privileged to provide a benchmark with real queries and real data. There are other data from real-world systems, such as UNIPROT (Universal Protein Resource)[8] which contains 380 million triples describing proteins.

---

[6] http://swat.cse.lehigh.edu/projects/lubm/.

[7] http://www.w3.org/wiki/Social_Network_Intelligence_BenchMark.

[8] http://dev.isb-sib.ch/projects/uniprot-rdf/.

In the performance column of Table 3.1, the performance results are coming from the implementer themselves. In this column, *BM* means the benchmark; *Max* means the maximum number of triples reported so far; *Load* means the speed of bulk loading in terms of the number of statements (either triples or quads) loaded in one second; *Query* is the query processing speed measured in terms of the number of statements returned in a second; and finally *Infer* means the inference speed (number of statements per second) and supported inference language profiles.

Related Storage Techniques

In addition to triple stores, there are some new trends in representing data or knowledge in non-relational database formats. The most popular buzz word in this regard might be *NoSQL*, which stands for "not only SQL", which reflects a recent move away from traditional relational databases. In that sense, triple stores as discussed above are a particular type of NoSQL database, because they are generally not based on relational techniques. Other types of NoSQL databases include document store (e.g. MongoDB), Key-Value Store, Object Database (e.g. DB4O) and wide columnar store (e.g. BigTable). Interested users can get more detailed information about NoSQL from its wikipage.[9]

Thinking of RDF databases from the angle of NoSQL databases, many readers might be interested in the question like *"How do RDF Databases Differ from Other NoSQL Solutions?"*. In general, the move away from relational databases was initiated by the adoption of key-value, document and wide columnar stores in order to improve scalability. These approaches share characteristics such as having *no or very lightweight schemas and being very scalable* as long as the **data is very heterogeneous** (e.g. you have large numbers of documents which have the same or similar structure and you have programs which know how to process those documents). However, these lightweight stores are not suitable for finding small pieces of information in the datastore, making links between documents or performing analytics, because they are optimised for simple storage and retrieval. This is where graph databases and triple stores fill the gap, because they are **flexible regarding the schema**: they can be used with no schema or with a lightweight schema, or they can be used with various schemas.

Besides this schema flexibility, the main advantages of RDF databases over other NoSQL stores are twofold. Firstly, they are **based on a series of W3C standards** and hence are designed based on the Web as a platform, rather than as standalone tools with custom interfaces. Standardisation in general also has advantages in terms of interoperability (avoids vendor lock-in and encourages competition between providers of RDF triple stores), data linkage, sharing, publishing, querying and many others. Secondly, RDF databases profit from the *Semantic Web* and *Linked Data* ecosystems, which have produced techniques, tools and libraries covering a wide spectrum of functionalities in data or knowledge-centric applications. For a detailed discussion in this regard, the interested readers can further check for an interesting

[9]http://en.wikipedia.org/wiki/NoSQL.

comparison between an RDF store and other NoSQL stores at http://blog.datagraph.org/2010/04/rdf-nosql-diff.

As an example of how the Linked Data ecosystem benefits storage techniques based on RDF, we discuss a new trend for improving the scalability of storage solutions. One of the drawbacks of RDF stores is that complex querying of large datasets can be computationally expensive to host. Hence, recently a new paradigm has been introduced called **Linked Data fragments** [241]. The main purpose of this paradigm is to deal with the server resource usage issues with most online SPARQL endpoints. The idea is to propose fragment types that require minimal server effort and *enable efficient client-side querying*. Although it is still unclear whether the approach in its current form can gain popularity in real-world Linked Data publishing, its underlying philosophy makes sense and future standards will likely include similar techniques to improve scalability and robustness of RDF data stores.

3.3.3 Property Graph-Based Stores

RDF Stores are not the only way to store knowledge graphs. There are alternative data models similar to RDF which can also be used. In this section, we briefly introduce property graphs, an alternative data model (not based on RDF) for representing knowledge graphs. Although in this book, we focus on RDF-based tools and methods, property graphs have also been widely adopted by graph database and graph-based analytics and application systems. As we will see, both data models are similar, thus the methods and algorithms described in this book can also be applied to knowledge graphs stored as property graphs.

A property graph is a graph in which nodes and edges can have multiple properties. In its simplest form, such properties can be described by key-value pairs. For example, Fig. 3.3 shows a property graph describing the relations among a project, a manager and a deliverable.

Interesting Features

A few *interesting features of property graphs* can be identified as given below:

- Both nodes and edges are first-class citizens of a property graph. Both can have properties. We call them the *elements* of the property graph.
- Property graphs can have directions. An undirected property graph can also be transformed into a directed property graph, by replacing each undirected edge with two directed edges with the same properties.
- Property graphs can be unlabelled. However, a labelled property graph can also be represented by an unlabelled property graph. Particularly, if a node $n \in N$ should have label $l(n)$, one can achieve this by adding a new key l into K, a new value $l(n)$ into V and a new property $(n, (l, l(n)))$ into P. Labels of edges can be represented in a similar manner.

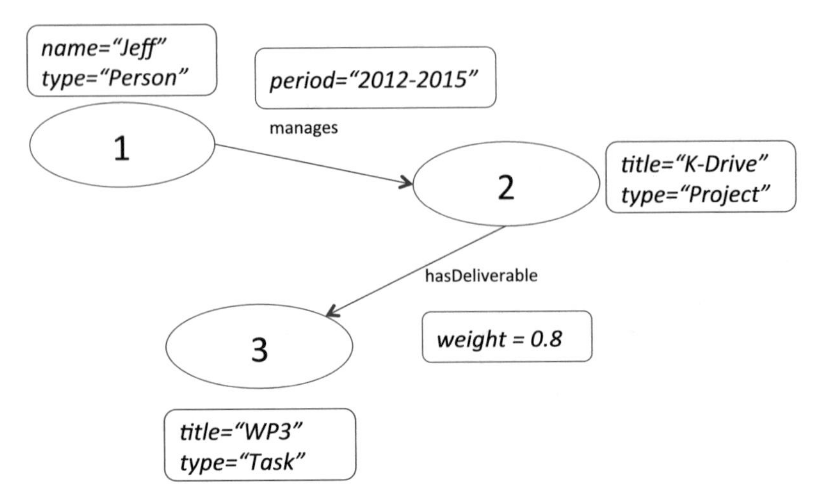

Fig. 3.3 Property graph example

- A property graph may have multiple key-value pairs with the same key for an element. This provides flexibility in real-world applications. For example, a person can have multiple names. Of course, a key can also be used to represent a unique property, such as an identifier.

Storage of Property Graphs

The features of property graphs discussed above suggest that, like RDF datasets, property graphs are very flexible data models and can be used to represent many other existing data models, which means that they can be stored in RDF triple stores, relational databases, NoSQL databases, etc.

For example, a schemaless RDF graph can be regarded as a property graph in which the edges cannot have any property other than labels, while a property of a node is also represented as a triple with the node as the subject, the key as the predicate and the value as the object.

Property graphs can also be transformed into many other data models:

- To represent a property graph with an RDF graph, one can represent edges in the property graph as nodes in the RDF graph, then all the properties can be represented with triples. Particularly, each attribute of a node or edge in the property graph can be represented as an RDF triple. This suggests that a property graph can always be serialised as an RDF document for data exchange.
- Property graphs can also be represented in a relational database. Particularly, the topology of the graph can be stored in a table with three columns, denoting the two nodes and the edge between them. The properties of the graph can also be stored in a table with three columns, denoting the element, the key and the value. This

suggests that property graphs can be persisted in and integrated with traditional Structured Query Language (SQL) database systems. Nevertheless, this way of persistence will lead to quite some redundancies in the database because the same elements will repeat multiple times when it has multiple properties. An alternative is to construct a table with each column denoting a property key and then populate the table with the property value for each element.
- It is also easy to represent a property graph with key-value collections. Both the properties of nodes and edges as well as the topology of property graphs can be represented using nested key-value pairs.[10] This suggests that property graphs can be operationalised with NoSQL databases and key-value computational frameworks. In fact, recent graph engines such as GraphX use key-value pair collection as the underlying representation of a property graph.

This section has discussed property graphs as an alternative data model and storage approach for a knowledge graph. We have seen that property graphs are roughly equivalent to RDF datasets, hence property graphs can be suitable to serve as the data model for knowledge graphs, in which many entities with different properties are connected.

3.3.4 Conclusion: Storing Knowledge Graphs Versus Relational Databases

As we have seen in this section, there are several approaches that can be used to provide access to—and to store—knowledge graphs, each with their own set of advantages and disadvantages. The OBDA approach is lightweight in the sense that it does not require the deployment of an additional storage database, but may not be suitable for intensive and exploratory use of the data. RDF and Property graph Store (or simply, graph database) can be acquired based on different data models, with varying degrees of standardisation, hence tools may need to be adapted in order to work with your chosen graph database provider.

One of the main considerations when choosing a graph database is about whether the same tasks can be performed with traditional relational databases (or simple NoSQL data stores). The key distinction is that, in the relational data model (and in simple NoSQL data models such as key-value, document and columnar stores), data are maintained and operated in collections, i.e. with tables, while in a property graph, data are maintained and operated in paths, i.e. with a sequence of connected nodes. Such a distinction leads to the following significant differences in practice:

---

[10]The properties can be represented with nested key-value pairs of form $(a, (k, v))$, where a, k and v are the element (node or edge in the property graph), key and value of the property, respectively. The topology of a property graph can also be represented with nested key-value pairs of form $(n_1, (e, n_2))$, where n_1, n_2 and e are the (identifiers of the) corresponding nodes and edge, respectively.

- In a relational data model, a schema is usually needed to describe the structure of the data in each table. This has both pros and cons. The benefit is that, when executing queries, the engine knows the structure of the data it is dealing with, by examining the schema of the table. While the drawback is that the schema has to be comprehensive enough to accommodate all data records in the table, even if many records do not have values for all columns. This can lead to a significant storage cost for the table and its indexes. At the same time, the schema has to be predefined for the indexing and querying engine to work in relational databases. This makes it less flexible and more difficult to add new properties to a data record or change the schema of an existing property because adding or modifying a column means extending or modifying all records in the table.
  RDF and Property Graph, by contrast, can be schemaless or schema-free. Extending the amount of data related to an element can easily be done without needing to adapt any schema. Any other nodes or edges in the graph will not be affected. RDF, in terms of flexibility, is even better than the Property Graph, because it also allows simple schema (with RDFS) or rich schema (with OWL). Unlike relational databases, schemas for RDF graphs are much easier to be updated and maintained. Hence, traditional relational databases (and basic NoSQL stores) are more convenient to represent large volumes of data with uniform, well-understood and persistent patterns. While RDF and Property Graph stores (or simply graph data stores) are more suitable to represent data that is varied, incomplete and/or dynamic. In scenarios such as knowledge graph, the property graph is useful due to its flexibility to support the discovery and modification of relations and attributes.
- In terms of accessibility if a query involves multiply connected nodes, then a relational database has to perform join among multiple tables to deliver the answers. The table-joining operation is known to be difficult when the number of tables is large and when the size of the table is large.
  In contrast to relational databases, graph databases can answer such queries by graph traversal. The graph database engine only needs to examine the nodes that are on the path of the query. Other nodes or edges in the property graph will not be involved. Since each record is processed independently, it reduces the performance and resource cost for query answering significantly. It also suggests that the processing of the graph can be distributed with respect to each node. This motivates the research and development of distributed graph computation.
  Therefore, graph data stores are more suitable than traditional relational databases when querying for paths. In this case, the expensive join operation in traditional databases can be avoided. Such a feature is particularly useful for applications where entities and queries are highly connected, for example, the Graph Search in Facebook.

In the next section, we will look at various ways of how knowledge graphs can be exploited, which will illustrate the advantages of graph databases over relational databases in some important use cases that are important for large organisations.

3.4 Knowledge Consumption Layer

The third main layer we consider in our architecture for knowledge graphs in organisations is the *knowledge consumption layer*. In this layer, you find various components which make it easier to access the knowledge stored in the graph by members of your organisation. In general, the rich structure and interconnectivity that can be achieved by storing information in knowledge graphs (rather than in traditional relational databases or document stores) facilitates the definition of intuitive components for consuming the data. In this section, we will introduce some of these consumption paradigms, such as *semantic search*, automatic *summarisations* and *question answering*. Although we do not explicitly mention reasoning here, reasoning is in fact required for supporting many of these tasks.

When considering the consumption of data we need to take into account the *types of consumers* related to your organisation which need to access the data. Naturally, consumers can be distinguished by their *role with regard to your organisation* (e.g. employees in particular business units, partners and customers) and you may want to restrict (or encourage) access to some parts of your knowledge graph. In general, authorisation issues can be dealt with in a similar fashion as with traditional databases or using existing authorisation infrastructures within your organisation; hence, we will not discuss these issues in this book.

Another way to distinguish data consumers is by their *technological knowledge* or skills. A minority of consumers within organisations will be technically savvy enough to be able to access knowledge directly from the database, e.g. using SPARQL queries. However, the vast majority of people in a typical organisation will require a more intuitive interface. The good news is that, by reusing standard vocabularies in your knowledge graph and because knowledge graphs can store metadata (e.g. syntactic information such as human-readable labels) together with the domain data, it is possible to create or reuse components which can analyse knowledge graphs to make their information accessible in easy-to-understand ways. Since this is one of the key advantages of knowledge graphs over other ways of data storage, we will focus on consumption paradigms which target mainly non-technological savvy users.

3.4.1 Semantic Search

One of the first and main applications of knowledge graphs which are widely known is that of *Semantic Search*. This is an extension of existing search approaches where knowledge about entities which are relevant to your organisation can be exploited to help users find documents and information more effectively.

Traditional (nonsemantic) search works in two phases:

- Indexing of documents: in this phase, documents (e.g. Web pages, reports, spreadsheets, emails) are ingested by the search application and **syntactically analysed**. Often, in order to improve the performance of this syntactic analysis, lists of synonyms need to be defined manually or extracted from existing databases and converted into formats which the search engine can process. The result of this phase is an index, where typically, keywords can be looked up quickly and linked to specific documents.
- Querying: in this phase, users enter keywords (or phrases) indicating what they are looking for. The search engine (syntactically) analyses these keywords and looks them up on the index, returning the matched documents.

Semantic search extends a traditional search engine using a knowledge graph in order to improve the analysis required during the indexing and querying phases:

- During indexing, the knowledge graph can be used to automatically generate a list of entities relevant to the organisation and their synonyms. Furthermore, the knowledge graph helps to identify and distinguish between ambiguous entities (a typical example is that "apple" can refer to either the computer company or the fruit). Finally, the knowledge graph can also help to categorise documents.
- During querying, advantages of the same analysis apply, which means that users can be given options to help them disambiguate their searches. Also, when presenting search results, summaries of the related entities (more on summarisations will be presented below) can be shown to help the users understand what the search engine has found.

In general, traditional (keyword-based non-semantic) search already works well in many cases. Semantic search further improves on this document retrieval paradigm by moving from *searching for (key)words* to *searching for things*. Most Web search engines (Google, Yahoo! Bing, Baidu) have recognised the value of these improvements and are already using knowledge graphs in order to provide this type of search.

3.4.2 Summarisation

Since knowledge graphs combine data from various sources and have an optional (flexible) schema, it is difficult to know what information they contain. In order to help users understand the contents of the knowledge graph (or put another way: to help knowledge workers in your organisation find out what information is available to your organisation), various summaries can be generated automatically by summarisation components.

Summaries can be generated at different levels of granularity and for different purposes. We give some examples:

- **Entity Summary** In this case, you can summarise an entity (a *thing* or *concept*) in your knowledge graph. These types of summaries help users to learn about the information contained in the knowledge graph and how entities are related to each other. Such summaries are well known from Web search engines, where they are displayed as *entity cards* (see, for example Fig. 3.4). Section 6.1 (p. 147) introduces the problem and approaches of entity summary in knowledge graphs.

Apple Inc.
Consumer electronics company
Apple Inc. is an American multinational corporation headquartered in Cupertino, California, that designs, develops, and sells consumer electronics, computer software, online services, and personal computers. Wikipedia
Customer service: 0844 209 0611
Stock price: AAPL (NASDAQ) US$118.93 -0.07 (-0.06%)
28 Nov, 13:07 GMT-5 - Disclaimer
Founded: April 1, 1976, Cupertino, California, United States
CEO: Tim Cook
Headquarters: Cupertino, CA, United States of America
Founders: Steve Jobs, Ronald Wayne, Steve Wozniak
Feedback

Fig. 3.4 Entity card as shown by Google when searching for "Apple"

- **Graph Summaries** At the opposite end of the spectrum from entity summaries, which only summarise one entity in the graph, you have summaries of the whole graph. In this case typically you will find various metrics about the graph, such as how many entities it has, or how many relations are there between entities. You may also find metadata such as where does the knowledge come from (data sources), when the graph was last updated, etc. Another useful information that is often included in graph summaries are the *main entities* in the graph. These are typically calculated by analysing the structure of the graph (e.g. entities which have many connections to other entities in the graph) or by taking into account information about how often people search for a particular entity. Section 6.2 on p. 154 discusses an Entity Description Pattern based approach for summarising knowledge graphs for the purpose of knowledge exploration.
- **Goal-driven Graph Profiling** While entity and graph summaries can be useful to get a feeling of the information contained in a knowledge graph, users often want to get more relevant summaries that are tailored to a particular task they are working on. In such cases, you can also generate summaries by first going through a step of eliciting which entities and relations are relevant for the particular task. For example, if a knowledge worker in your organisation is working for a large client organisation, your knowledge graph may contain many entities related to that client. By specifying for example, which business unit within the client organisation is relevant or which areas are relevant for the current project, the knowledge worker can get a custom view of the entities in the knowledge graph which are more relevant for the task at hand. Interested readers can refer to Sect. 6.3 starting

from p. 155 for more detailed information about how to carry out knowledge graph profiling based on the given technical goals.

- **Graph Analytics** Finally, there are also many analytics which you can extract from knowledge graphs. Many of these analytics overlap with or are related to the goal-driven and the generic graph summaries, but in this case we are not interested in specific tasks or in giving generic information about the knowledge graph. In graph analytics, we are only interested in finding *interesting* patterns in the data, such as, for example: which queries can be answered by the knowledge graph, how are entities of particular types related in the graph or how do particular entities change over time. Typically, these types of graph analytics cannot be fully automated, but need to be performed semi-automatically.

In Chap. 6, we will look at various summarisation types for knowledge graphs in more detail.

3.4.3 Query Generation

One of the main challenges in exploiting knowledge graphs is to help users understand the usefulness of its knowledge or data, in terms of what kinds of queries can be answered, without requiring the users to be aware of the complexity of the underlying data model.

Query answering using structured query languages, e.g. SPARQL, is an important means to exploit knowledge graphs. There has been an abundance of research on answering semantic queries, but only limited attention has been paid to constructing queries that can most effectively retrieve the insightful information in a dataset.

Indeed, query generation is a non-trivial problem in knowledge graph exploitation. From the users' perspective, they are not always familiar with underlying data/knowledge technologies such as graph data model, RDF or SPARQL. They are also not always familiar with the datasets they are dealing with, especially when the datasets are of large scale and are linked by/to many remote datasets. From a knowledge representation's perspective, queries are important means for describing the contents of the datasets as their solutions correspond to subsets of the datasets that satisfy certain constraints. In this regard, an insightful query usually reveals some meta-knowledge, such as topics, trends, of knowledge graphs.

Query generation (QG) has been studied in the field of database with the main motivation of testing databases. Some QG approaches (such as [217]) are based on database schemas, while others (such as [162]) are based on actual data in databases. A related research problem is query recommendation (QR), where query logs are widely used to generate queries based on the querying and browsing behaviours of users. These approaches (such as [48]) rely on the knowledge about users. Similar approaches (such as [259]) are also used in information retrieval.

There has been some work on QG in the field of Semantic Web. Similar to the work in database, most of the existing work is for testing Semantic Web engines.

For instance, Cuenca Grau and Stoilos [93] presented an approach to generating queries to test incomplete ontology reasoners for EL, DL-Lite and DLP. Their generated queries are based on the ontologies rather than actual data. Görlitz et al. [92] proposed to generate queries for testing Linked Data query engines, based on some *input parameters*.

Nevertheless, both these approaches are designed to generate queries for evaluating the quality or performance of implementations, instead of assisting users in the exploitation of semantic data. As far as we know, there is no existing work on the systematic generation of semantic queries, based on the analysis of given semantic data, for the purpose of facilitating users to understand the data with insightful queries.

In terms of query generation for knowledge graphs, it is more interesting and important to unlock the ability of revealing interesting information in the graph to users. As a consequence, it should focus on the characteristics and insights of the knowledge graphs, as well as people's requirements to queries.

In the context of this book, query generation facilitates the exploitation of large-scale knowledge graphs, helping users to understand the contents of graphs and to identify the most relevant ones. The work presented in Sect. 6.4 starting from p. 169 will be the foundation for future research on query generation and data exploitation.

3.4.4 *Question Answering*

Traditional (keyword-based) search and semantic search are currently the main ways that knowledge workers use to access organisational information. However, the *search paradigm* only covers part of the process that knowledge workers perform. Typically, a knowledge worker will have a particular question in mind, which can be translated as a list of keywords to search for. The main disadvantage of this approach is that it *returns a list of documents* that match the searched keywords. After the search query, the knowledge worker still needs to open the found documents and extract the information they were looking for (if they are in fact in the found documents). *Question Answering* aims to automate this process whenever possible. The main idea here is that instead of translating the question into keywords and then reading through the found documents, the knowledge worker can directly ask the question and get an answer if the information is available.

In order for question answering to work, the system must be able to analyse incoming natural language questions (written or spoken) and correctly map these into queries to the knowledge graph (or a semantic search engine). Depending on the question and whether the information is available directly from the knowledge graph or as part of a previously analysed document, the system then needs to determine whether it can answer the question and how to present the answer to the user.

Historically, question answering has been attempted on top of relational databases. However, since relational databases have a fixed schema, have a specific purpose (to serve as the backend of a particular application) and do not contain additional metadata (such as names or synonyms of columns), the resulting question answering systems require the definition of additional information to extend the database schema and they can only answer a limited number of questions. The flexible schemas in knowledge graphs and the common vocabularies for adding names and synonyms to entities in the knowledge graph mean that generic question answering components, which can interpret questions and find answers in a given knowledge graph, can be developed.

In Chap. 7 we will present the various types of question answering systems in more detail as well as describe generic architectures for these question answering systems. Here, we will limit ourselves to presenting some current examples of what is currently possible with question answering systems over knowledge graphs.

One of the best known examples of question answering is the Watson computer which competed against the best human candidates in the Jeopardy TV show in 2011 [77]. Watson combines techniques for textual question answering with question answering over knowledge graphs. Watson has a very complex architecture and requires customisation and supercomputer infrastructure in order to adapt to new areas, but indicates what will be possible in the future.

Another example of question answering over knowledge graphs is that many Web search engines already provide an answer to some questions. For example, in Figs. 3.5 and 3.6, we can see that Google provides a direct answer to questions "who is the CEO of apple" and "how old is Tim Cook?". At the time of writing this book, Google is being conservative in the types of questions it answers. For example, Google does not answer "How old is the CEO of Apple?", even though it could do so based on the answers to the earlier questions. In Chap. 7 we will see that some QA systems can find answers to such questions.

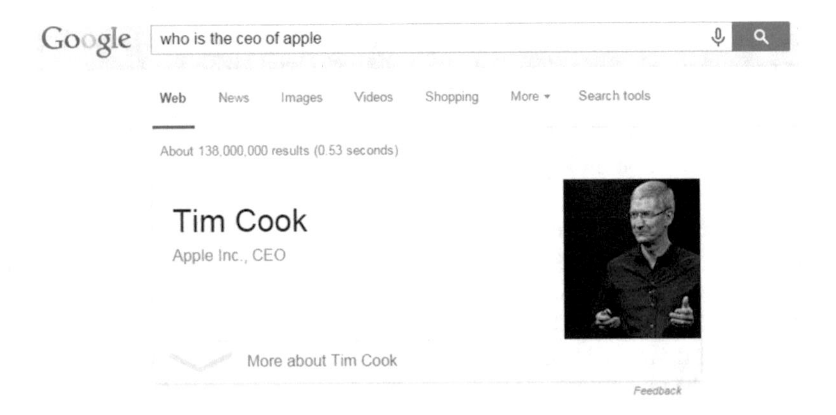

Fig. 3.5 Question Answering example by the Google Search Engine

Fig. 3.6 Another Question Answering example by the Google Search Engine

A final example of question answering over knowledge graphs is the Facebook Graph Search.[11] Although specific technical details of this system have not been published at the time of writing this book, in this case, Facebook seems to have defined a controlled natural language for posing questions based on their Facebook Graph (their version of a knowledge graph). The language they defined allows them to interpret questions posed by users and map these to queries of their graph. Example questions are:

- "Photos of my friends in New York",
- "Restaurants in London my friends have been to" or
- "Cities my family visited".

3.4.5 Conclusion

In this section, we looked at various components which can take advantage of your knowledge graph in order to make the information available to users related to your organisation. As we saw, knowledge graphs enable the improvement of current Information Retrieval paradigms such as keyword-based search engines by providing *semantic search* capabilities. Knowledge graphs also enable the exploration and understanding of your organisational information by providing summaries at different levels of granularity (from single entities to entire knowledge graphs) and taking into account different goals. Finally, we saw that knowledge graphs can play a central role in moving from a search paradigm for finding information to *Question Answering* where users can get direct answers to particular questions.

In this section, we have focused on components that facilitate the consumption of information by end users (i.e. users who are not technologically savvy), which is one of the key differentiators between knowledge graphs and traditional databases or

[11] https://en-gb.facebook.com/about/graphsearch.

document stores. However, more technical interfaces to directly access the knowledge graphs are always an option. For example, by providing a SPARQL endpoint to your knowledge graph, any developer can start querying the knowledge graph to perform analytics or to present information to end users. Also, RESTful services provide low-level services such as entity search and disambiguation, document enrichment or document classification, etc.

Part II
Constructing, Understanding and Consuming Knowledge Graphs

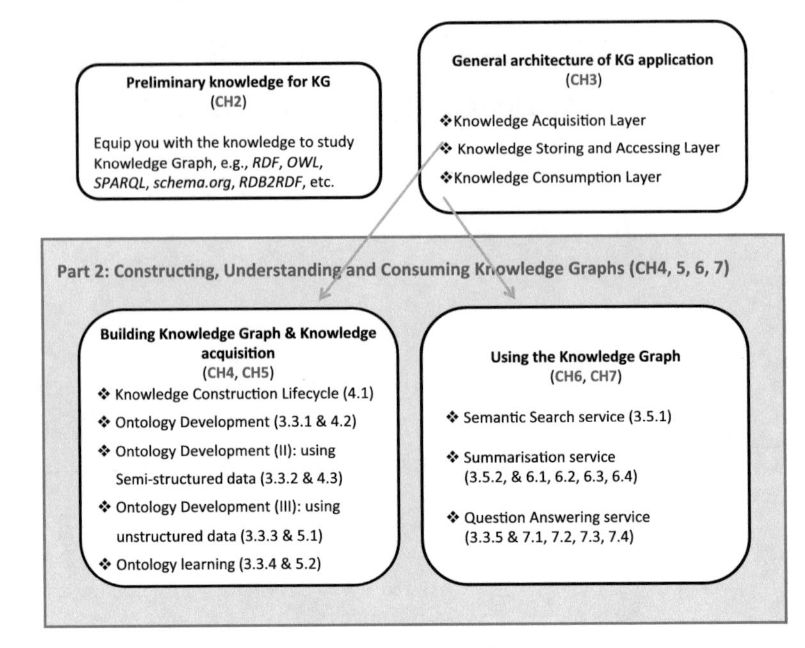

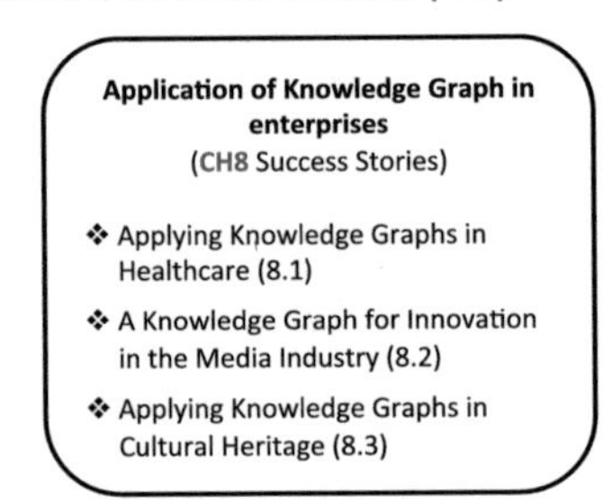

Application of Knowledge Graph in enterprises
(CH8 Success Stories)
❖ Applying Knowledge Graphs in Healthcare (8.1)
❖ A Knowledge Graph for Innovation in the Media Industry (8.2)
❖ Applying Knowledge Graphs in Cultural Heritage (8.3)

Fig. RoadMap. 2 The roadmap of Part II

The second part of this book focuses on the key technologies in building and using Knowledge Graphs. It is organised under the first and third layers in general architecture of Knowledge Graph Fig. 3.1:

I. *Acquisition* and *Integration Layer*
  Chapter 4: Construction of Enterprise Knowledge Graphs (I)
  Chapter 5: Construction of Enterprise Knowledge Graphs (II)
II. *Knowledge Storage layer* (Section 3.3)
III. *Knowledge Consumption layer*
  Chapter 6: Understanding Knowledge Graphs
  Chapter 7: Question Answering and Knowledge Graphs

Chapter 4
Construction of Enterprise Knowledge Graphs (I)

Boris Villazon-Terrazas, Nuria Garcia-Santa, Yuan Ren, Kavitha Srinivas, Mariano Rodriguez-Muro, Panos Alexopoulos and Jeff Z. Pan

In the previous chapters, we have shown the three-layer architecture of Knowledge Graph for Organisations. The very first question we are facing is how to build the knowledge graphs. From now on, we will use two chapters to introduce the technologies for the *Acquisition and Integration Layer*. In particular, we start with a generic lifecycle for constructing and maintaining knowledge for knowledge graphs in Sect. 4.1 and, then, we elaborate on the knowledge graph construction approaches a.k.a. the modelling (Ontology Authoring) and data lifting (Semantic Tagging and Interlinking) steps in the lifecycle.

In the current chapter, we focus on supervised approaches to constructing *new* knowledge, i.e. approaches involving human effort. Specifically, for the modelling step we introduce a competency question based ontology authoring framework, while

B. Villazon-Terrazas (✉) · N. Garcia-Santa · P. Alexopoulos
Expert System, Prof. Waksman 10, 28036 Madrid, Spain
e-mail: bvillazon@expertsystem.com

N. Garcia-Santa
e-mail: ngarcia@expertsystem.com

P. Alexopoulos
e-mail: palexopoulos@expertsystem.com

Y. Ren · J.Z. Pan
University of Aberdeen, King's College, Aberdeen AB24 3UE, UK
e-mail: y.ren@abdn.ac.uk

J.Z. Pan
e-mail: jeff.z.pan@abdn.ac.uk

K. Srinivas · M. Rodriguez-Muro
IBM USA, Thomas J. Watson Research Center, Yorktown Heights, NY 10598, USA
e-mail: ksrinivs@us.ibm.com

M. Rodriguez-Muro
e-mail: mrodrig@us.ibm.com

J.Z. Pan et al. (eds.), *Exploiting Linked Data and Knowledge Graphs in Large Organizations*, DOI 10.1007/978-3-319-45654-6_4

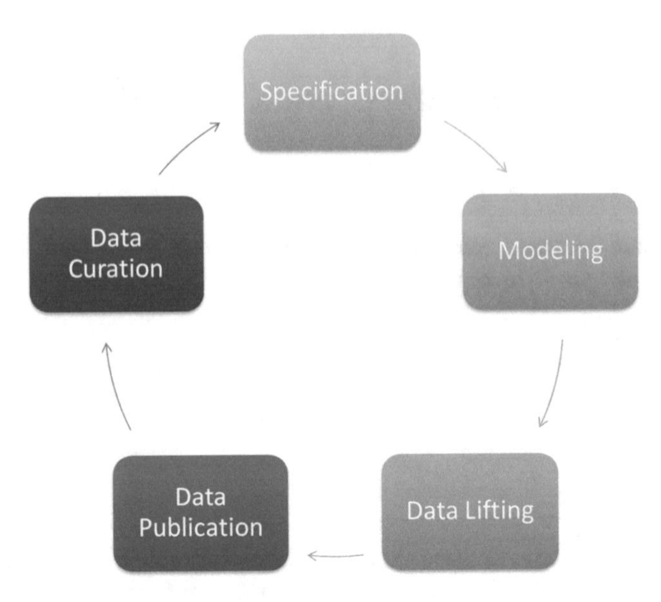

Fig. 4.1 Knowledge construction and maintenance lifecycle

for data lifting we discuss a semi-automated approach for creating linkages among heterogeneous data sources. In the next chapter, we put more focus on automated construction approaches.

4.1 Knowledge Construction and Maintenance Lifecycle

As we mentioned in Chap. 2 the information in a knowledge graph is derived from many sources within enterprises which need to be extracted, integrated and extended under a common schema or vocabulary. The effort required to do this is not trivial and is very similar to the effort required to implement *answer engines*, such as IBM Watson or the publishing of existing data as Linked Data (e.g. converting the unstructured knowledge in Wikipedia as DBpedia). Therefore, it is important to think about how best to organise the various tasks that need to be performed during the creation (and exploitation) of knowledge graphs.

This section describes a general pipeline for generating, publishing and evolving knowledge graphs. The pipeline follows an iterative and incremental lifecycle, based on an existing Linked Data lifecycle model that has been already applied in real case scenarios [122, 245]. The activities of the pipeline, as depicted in Fig. 4.1, include: (1) specification, (2) modelling, (3) data lifting, (4) data publication and (5) data curation.

(1) Specification

The first step to start working in a knowledge graph construction process is the drawing up of a detailed specification of requirements. It has been proved that detailed requirements provide several benefits, some of which are:

- The establishment of the basis for agreement between customers and developers on what the application is supposed to do.
- The provision of a basis for estimating costs and schedules.
- The offer of a baseline for validation and verification.
- The reduction of the development effort.

Apart from the basic requirements, domain-specific requirements are also important to depict properly the special features for each data and work area. So, these requirements should be identified in addition to the application ones. At this stage, the description of this activity is not intended to be exhaustive but it just introduces the most important points. For this, the main tasks identified are (1) identification and analysis of the data sources, and (2) URI design.

Identification and analysis of the data sources

We need to identify and select the data that we want to integrate and publish within our knowledge graph. We have to distinguish between the data that is already available inside the organisation and the data that is not. In the first case, the organisation needs merely to gather the different data sources whereas, in the second case, it needs to ensure the timely acquisition of the external data, e.g. by purchasing it.

After we have identified and selected the data sources, we have to (i) search and compile all the available data and documentation about those resources, including purpose, components, data model and implementation details; (ii) identify the schema of those resources including the conceptual components and their relationships; and (iii) identify real-world entities and their relationships which are included in the data sources [109].

URI design

To achieve a globally interlinked Knowledge Graph it is necessary to identify a resource on the Internet or in a local intranet within an organisation; for that purpose URIs are used. The URIs should be designed with simplicity, stability and manageability in mind, thinking of them as identifiers rather than as names for Web resources. Some guidelines for designing URIs are as follows [205]:

- Use meaningful URIs, instead of opaque URIs, when possible. It is recommended to put into the URI as much information as possible.
- Use slash (303) URIs, instead of hash URIs, when possible.

- Separate the TBox (ontology model) from the ABox (instances) URIs. To do that, we have to manage the following URI elements:
  - Base URI structure. Here we need to choose the right domain for URIs.
  - TBox URIs. We recommend to append the word *ontology* to the base URI structure. Then, we would append all the ontology elements, classes and properties.
  - ABox URIs. We recommend to append the word *resource* to the base URI structure. Additionally, we recommend to use Patterned URIs by adding the class name to the ABox base URI.

(2) Modelling

After the specification activity, in which the data sources were identified, selected and analysed, we need to determine the ontology to be used for modelling the domain of these data sources. The most important recommendation in this context is to reuse as much as possible the available vocabularies. This reuse-based approach speeds up the ontology development, saving time, effort and resources [31]. This activity consists of the following tasks:

- Search for suitable vocabularies to reuse [225]. Currently, there are a few useful repositories to find available vocabularies, such as Swoogle[1] and LOV.[2]
- In the case that we cannot find any vocabulary that is suitable for our purposes, we should create it, trying to reuse as much as possible the existing resources, e.g. resources available at sites like http://semic.eu/, etc. [244].
- Finally, if we are not able to find the available vocabularies or resources for building the ontology, we have to create the ontology from scratch. To this end, we can follow the first scenario proposed in the NeOn Methodology [224].

(3) Data Lifting

The Resource Description Framework, RDF,[3] is the standard data model for data interchange and the format for our knowledge graph. Therefore, in this activity we have to take the data sources selected in the specification activity and transform them to RDF according to the vocabulary created in the modelling activity (see section "Modelling"). The data lifting activity consists of transformation and linking.

Transformation

The preliminary guidelines proposed in this chapter consider only the transformation of the whole data source content into RDF, i.e. following an Extract, Transform and Load ETL-like,[4] using a set of RDF-isers, i.e. RDF converters. The requirements

---

[1]http://swoogle.umbc.edu/.

[2]Linked Open Vocabularies http://lov.okfn.org/.

[3]http://www.w3.org/RDF/.

[4]Extract, transform and load (ETL) of legacy data sources is a process that involves: (1) extracting data from the outside resources, (2) transforming data to fit operational needs and (3) loading data into the end target resource process [132].

of the transformation are (1) full conversion, which implies that all queries that are possible on the original source should also be possible on the RDF version; and (2) the RDF instances generated should reflect the target ontology structure as closely as possible, in other words, the RDF instances must conform to the already available ontology/vocabulary.

There are several tools that provide technological support to this task. In Sects. 2.3.1 and 2.3.2, we have introduced RDB2RDF and GRDDL, respectively. Here, we provide a list of tools with a very brief description.[5]

- For CSV (comma-separated values) and spreadsheets: RDF extension of Google Refine,[6] XLWrap,[7] RDF123[8] and NOR_2O.[9]
- For relational databases: D2R Server,[10] ODEMapster,[11] Triplify,[12] Virtuoso RDF View[13] and Ultrawrap.[14] It is worth mentioning that the RDB2RDF Working Group[15] releases R2RML,[16] a standard language to express mappings between relational databases and RDF.
- For XML: GRDDL[17] through XSLT, TopBraid Composer and ReDeFer.[18]
- For other formats any23[19] and Stats2RDF.[20]

Linking

Following the fourth Linked Data Principle ("*Include links to other URIs, so that they can discover more things*"), the next task is to create links between our knowledge graph and optionally external knowledge graphs. This task involves the discovery of relationships between data items. We can create these links manually, which is a time-consuming task, or we can rely on automatic or supervised tools. The task consists of the following steps:

---

[5].For a complete list see http://www.w3.org/wiki/ConverterToRdf.

[6]http://lab.linkeddata.deri.ie/2010/grefine-rdf-extension/.

[7]http://xlwrap.sourceforge.net/.

[8]http://rdf123.umbc.edu/.

[9]http://www.oeg-upm.net/index.php/en/downloads/57-nor2o.

[10]http://sites.wiwiss.fu-berlin.de/suhl/bizer/d2r-server/.

[11]http://www.oeg-upm.net/index.php/en/downloads/9-r2o-odemapster.

[12]http://triplify.org/.

[13]http://virtuoso.openlinksw.com/whitepapers/relational%20rdf%20views%20mapping.html.

[14]http://www.cs.utexas.edu/~miranker/studentWeb/UltrawrapHomePage.html.

[15]http://www.w3.org/2001/sw/rdb2rdf/.

[16]http://www.w3.org/TR/r2rml/.

[17]http://www.w3.org/TR/grddl/.

[18]http://rhizomik.net/redefer/.

[19]http://any23.org/.

[20]http://aksw.org/Projects/Stats2RDF.

- Identify knowledge graphs that may be suitable as linking targets. For this purpose we can look for knowledge graphs of similar topics on the Linked Data repositories like CKAN (Comprehensive Kerbal Archive Network).[21] Currently, there is no tool support for this, so we have to perform the search in the repositories manually. However, there are approaches to perform this step, such as [153, 180].
- Discover relationships between data items of our knowledge graph and the items of the identified knowledge graphs in the previous step. There are several tools for creating links between data items of different knowledge graphs, for example the SILK framework [35] or LIMES [175].
- Validate the relationships which have been discovered in the previous step. This is usually performed by domain experts.

(4) Data Publication

In this step, we review the publication of RDF data. In a nutshell, this activity consists of (1) knowledge graph publication and (2) metadata publication. These activities are described next.

Knowledge graph publication

Once we have the legacy data transformed into RDF, we need to store and publish that data in a triple store.[22] There are several tools for storing RDF datasets, for example Virtuoso Universal Server,[23] Jena,[24] Sesame,[25] 4Store,[26] YARS[27] and OWLIM.[28] Some of them already include a SPARQL endpoint and Linked Data front-end. However, there are some tools like Pubby,[29] Joseki[30] and Talis Platform[31] that provide these functionalities. A good overview of the recipes for publishing RDF data can be found in [109].

Metadata Publication

Once our knowledge graph is published we have to include the metadata information about it. For this purpose there are vocabularies such as (1) VoID[32] that allows to express metadata about RDF datasets, and it covers general metadata, access

[21] http://ckan.net/group/lodcloud.

[22] A triple store is a purpose-built database for the storage and retrieval of RDF.

[23] http://virtuoso.openlinksw.com/.

[24] http://jena.sourceforge.net/.

[25] http://www.openrdf.org/.

[26] http://4store.org/.

[27] http://sw.deri.org/2004/06/yars/.

[28] http://www.ontotext.com/owlim.

[29] http://www4.wiwiss.fu-berlin.de/pubby/.

[30] http://www.joseki.org/.

[31] http://www.talis.com/platform/.

[32] http://www.w3.org/TR/void/.

metadata, structural metadata and description of links between knowledge graphs; and (2) Open Provenance Model[33] that is a domain-independent provenance model result of the Provenance Challenge Series.[34]

(5) Data Curation

This activity aims at cleaning, maintaining and preserving data for reuse over time. The paradigm of generating, publishing and exploiting knowledge graphs has inevitably led to several problems. There is a lot of noise which inhibits applications from effectively exploiting the structured information that underlies a knowledge graph [118]. This activity focuses on cleaning this noise, e.g. the linked broken data. It consists of two steps:

- Identify and find possible mistakes. To this end, Hogan et al. [118] identified a set of common errors:
  - http-level issues such as accessibility and derefencability, e.g. HTTP URIs return 40x/50x errors.
  - reasoning issues such as namespace without vocabulary, e.g. `rss:item`; a term invented in the related namespace, e.g. `foaf:tagLine` invented by LiveJournal; the term is a misspelt version of the term defined in namespace, e.g. `foaf:image` versus `foaf:img`.
  - malformed/incompatible datatypes, e.g. "true" as `xsd:int`.
- Fix the identified errors. For this purpose Hogan et al. [118] also propose some solutions at the application side and the publishing side.

4.2 Ontology Authoring: A Competency Question-Driven Approach

Many real-world ontologies are constructed manually by human authors. Such manual ontology authoring is crucial for ensuring the high quality of the knowledge system, especially in domains, e.g. biomedicine, where deep and complex professional knowledge needs to be represented. Nevertheless, ontology authoring remains a challenging task. Studies on ontology authoring, such as experiences from the OWL Pizzas tutorial [199] and the NeOn project [72], suggest that ontology formalisms are often not straightforwardly comprehensible and logical implications can be difficult to resolve. This is because ontology authors are usually domain experts but not necessarily proficient in ontology technologies, especially their logic underpinnings. As a consequence, on the one hand it is difficult for human authors to express their requirements for the axiomatisation of an ontology and, on the other hand, it is also difficult to know whether the requirements are fulfilled as a result of their ontology

[33]http://openprovenance.org/.

[34]http://twiki.ipaw.info/bin/view/Challenge/OPM.

authoring actions. Hence, ontology authoring is usually time consuming, error-prone and requires extensive training and experience [199].

To address this issue, in this section we introduce the methodology of Competency Question-driven Ontology Authoring (CQOA) [200],[35] which leverages the ideas of *competency questions* and *testing driven software development* (where a suite of tests represent a specification for a program and the tests are coded against).

4.2.1 Competency Questions

(Informal) Competency Questions (CQs) are expressions of questions that an ontology must be able to answer [237]. We consider these to be natural language sentences that express patterns for types of question people want to be able to answer with the ontology. The ability to answer questions of the type indicated by a CQ meaningfully can be regarded as a *functional requirement* that must be satisfied by the ontology.

Example 1 Below are some example CQs:

- "Which mammals eat grass?" (in an animal ontology)
- "Which processes implement an algorithm?" (in a software engineering ontology)

Compared to more formal requirement specifications, CQs are particularly useful to ontology authors who are less familiar with description logics because CQs are in natural language, are about domain knowledge and do not require an understanding of DLs. Hence in ontology authoring practice, CQs help authors to determine the scope and granularity of the ontology, and to identify the most important classes, properties and their relations.

We have two important observations on CQs:

1. Following on from the ideas of Frege, many philosophers of language use the term *presupposition* to refer to a special condition that must be met for a linguistic expression to have a denotation [22]. The fact that a question may have presuppositions, and that these may represent misconceptions on the part of the asker, has been exploited by researchers working on principles for cooperative question answering from databases [83]. From a linguistic point of view, CQs also have presuppositions about the domain of discourse that have to be satisfied to ensure that the answers to CQs are cooperative:

   Example 2 In order to meaningfully answer the CQ "Which processes implement an algorithm?" it is necessary for the ontology to satisfy the following presuppositions:

[35] We are grateful that our co-authors of [200] allow us to reuse some of the content here in this section from our joint paper.

(a) Classes *Process*, *Algorithm* and property *implements* occur in the ontology;
(b) The ontology allows the possibility of *Process*es implementing *Algorithms*;
(c) The ontology allows the possibility of *Process*es *not* implementing *Algorithm*s.

The last two of these perhaps need some justification. If case 2 was not satisfied, the answer to all *Process*es and all *Algorithm*s would be "none", because the ontology could never have a *Process* implementing an *Algorithm*. This would exactly be the kind of uncooperative answer looked at by the previous work on cooperative question answering [83]. It is hard to imagine an ontology author really wanting to retrieve this information. Rather, this can be taken as evidence of possible design problems in the ontology. If case 3 was not satisfied, the answer to all the *Algorithm*s would be a list of all the *Process*es. This would mean that the questions would be similarly uninteresting to the ontology author, again signalling a possible problem in the ontology.

2. CQs can have clear and relatively simple syntactic patterns. For example, the CQs in Example 1 are all of the following semiformal pattern:

$$Which\ [CE1]\ [OPE]\ [CE2]?$$

where $CE1$ and $CE2$ are class expressions (or individual expressions as a special case) and OPE is an object property expression. This pattern asks for instances or subclasses of $CE1$ that can have an OPE relation to some instance of $CE2$. With such patterns, the presuppositions shown in Example 2 can be verified automatically:

(a) $CE1$, $CE2$ and OPE should occur in the ontology;
(b) $CE1 \sqcap \exists OPE.CE2$ should be satisfiable in the ontology, where $CE1 \sqcap \exists OPE$.
$CE2$ is the DL formula for "CE1 that has some OPE to CE2";
(c) $CE1 \sqcap \neg(\exists OPE.CE2)$ should be satisfiable in the ontology. Here $\neg$ is the constructor for negation.

We call tests of this kind which can be derived from CQs' *Authoring Tests* (ATs).

The idea of CQOA is to support the ontology author in the formulation of machine-processable CQs for their ontology. In an implemented system, the users will be allowed to either import their predefined CQs or enter new CQs in a controlled natural language. The authoring environment will identify the patterns of the inputted CQs and generate appropriate ATs. With the ATs, certain aspects of the answerability of the CQs can then be tested by the authoring environment to find places where the ontology does not yet meet the requirements. If there is a change in the status of these ATs from true to false or vice versa, the system will report the result to the users. The pattern identification, AT generation and testing procedures are all transparent to authors hence they can be utilised by novice ontology authors.

4.2.2 Formulation of Competency Questions

To realise this CQOA workflow, we first need to understand how real-world CQs are formulated. From the second observation of CQ, one can realise that CQs usually consist of certain elements, such as class expressions, object property expressions, and these elements are used with clear patterns. In order to represent the commonality and variability of different CQ patterns, we employed the feature-based modelling method [183]. Based on the analysis of different CQ patterns observed in real-world CQ collections that are constructed by actual ontology authors and users, the feature hierarchy in Fig. 4.2 can be identified:

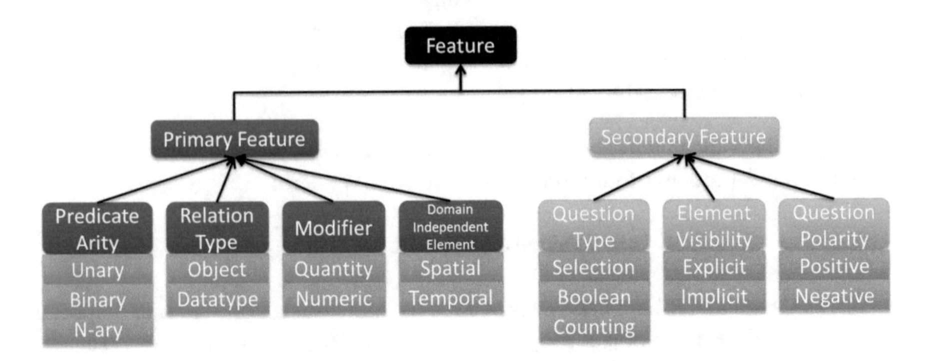

Fig. 4.2 CQ feature hierarchy

In this hierarchy, each feature has several variabilites. For example, the variabilities of *Question Type* are *Selection Question*, *Binary Question* and *Counting Question*. The variabilities of *Primary Features* affect the required modelling elements of the ontology. The variabilities of *Secondary Features* do not. We introduce each feature and its variabilities as follows:

1. **Question Type** determines the kinds of answers presented when answering the CQ:
   (a) *Selection question* should be answered with a set of entities or values that satisfy certain constraints. The CQs in Example 1 are all selection questions.
   (b) *Binary question* should be answered with a boolean value, i.e. *yes* or *no*, indicating the existence of any answer to a selection. For example, "Does this pizza contain halal meat?" is a binary question corresponding to a selection question "Which of these pizzas contain halal meat?".
   (c) *Counting question* should be answered with the number of different answers to a selection question. For example, "How many pizzas have either ham or chicken topping?" is a counting question. Its corresponding selection question is "Which pizzas have either ham or chicken topping?".

2. **Element Visibility** indicates whether the modelling elements, such as the class expressions and property expressions, are *explicit* or *implicit* in the CQ. For example, "What are the export options for this software?" has explicit elements *Software* and *Export Option*, but also an implicit relation *hasExportOption* between softwares and export options. Note that even implicit elements should occur in the ontology to make the CQ meaningful.
3. **Question Polarity** determines if the question is asked in a *positive* or *negative* manner, e.g. "Which pizzas contain pork?" versus "Which pizza has no vegetables?"
4. **Predicate Arity** indicates the number of arguments of the main predicate:

    (a) *Unary predicate* is concerned with a single set of entities/values and its instances, e.g. "Is it thin or thick bread?"
    (b) *Binary predicate* is concerned with the relation between two sets of entities/values and their instances, such as the *eat* and *implement* in Example 1.
    (c) *N-ary predicate* is concerned with the relation among multiple ($\geq$3) sets of entities/values and their instances. Given the fact that DLs can only represent unary and binary predicates, an N-ary predicate has to be represented as a concept via reification. In the next section, we will show how this affects the ATs.

5. **Relation Type** indicates the kind of relation for the main relation involved in the CQ. As in DLs, CQs can have object property relations or datatype property relations. Note that a relation with more than two arguments or with its attributes has to be represented by an entity via reification.
6. **Modifier** is employed to impose restrictions on some entities/values:

    (a) *Quantity modifier* restricts the number of relations among entities/values.

        i. It can be a concrete value or value range. For example, "If I have 3 ingredients, how many kinds of pizza I would make?" has a quantity modifier 3 on the number of pizza–ingredient relations for each pizza.
        ii. It can be a superlative value or value range. For example, "Which pizza has the most toppings?" has a quantity modifier *most* on the number of pizza-topping relations for each pizza.
        iii. It can also be a comparative value or value range. For example, "Which pizza has more meat than vegetables?" has a quantity modifier *more* on the number of pizza–meat and pizza–vegetable relations for each pizza.

    (b) *Numeric modifier* is used to restrict the value of some datatype properties. Similarly to the quantity modifier, it can be a concrete value or value/range, or a superlative value, or a comparative value. For example, "What pizza has very little ($\leq$10 %) onion and/or leeks and/or green peppers?".

7. **Domain-independent Element** is an element that can occur across different knowledge domains. It is usually associated with some physical or cognitive measurements. Some most commonly used domain-independent elements include:

(a) *Temporal element* in the CQ indicates that the CQ is about the time of some event, e.g. "When was the 1.0 version released?".
(b) *Spatial element* in the CQ indicates that the CQ is about the location of some event. It does not have to be a physical location. For example,"Where is the documentation?" can be answered with a file path or a URL.

CQs with different primary features are distinguished into different archetypes. CQs with different secondary features in an archetype are distinguished into different subtypes. Together, they constitute a generic framework to formulate different CQ patterns. For example, the CQ pattern *Which [CE1] [OPE] [CE2]*? features a selection question with binary predicate of an object property relation and all elements are explicit. With the feature-based framework, different CQ archetypes and subtypes can be constructed. Table 4.1 illustrates 12 CQ archetypes that are observed in real-world scenarios, in which 1, 2, 3, 6, 8, 10 are more frequently used [200]. The first column shows the ID of the archetype, the second and third columns show the pattern and one example from our collection. The last four columns are the primary features. As we mentioned above, some archetype patterns have subtypes. An example of the subtypes of archetype 1 is illustrated with Table 4.2, in which the last three columns are the secondary features.

The feature-based framework is also flexible enough to describe more complex CQs. For example, a hypothetical CQ "How many pieces of software are most efficient when providing this service?" has a pattern *How many [CE1] are [NM] to [OPE] [CE2]*?, which is a counting question subtype in archetype 6.

4.2.3 Ontology Authoring Workflow

To realise the CQOA workflow, we now need to automatically test whether a CQ can be meaningfully answered. As shown before, this can be achieved by generating AT test suites based on the elements and features of CQs. Given that our framework describes the CQs in terms of a set of features, we first analyse the presuppositions implied by different variations of each feature:

1. **Question Type:** regardless of the question type, the modelling elements mentioned in the question should **occur** in the ontology. Classes should also be **satisfiable**.

   (a) *Selection question* asks for the answers satisfying certain constraints. The ontology should allow some answers to satisfy the constraints. For example, "Which pizzas contain pork?" implies that pork is allowed to be contained in pizzas, i.e. $Pizza \sqcap \exists contains.Pork$ should be **satisfiable**. Otherwise, no pizza can contain pork at all. The ontology should also allow some entities to NOT satisfy the constraints. For example, the CQ above implies that it is possible for some pizza to contain no pork, i.e. $Pizza \sqcap \forall contains.\neg Pork$ is

Table 4.1 CQ Archetypes (PA = Predicate Arity, RT = Relation Type, M = Modifier, DE = Domain-independent Element; obj. = object property relation, data. = datatype property relation, num. = numeric modifier, quan. = quantitative modifier, tem. = temporal element, spa. = spatial element; CE = class expression, OPE = object property expression, DP = datatype property, I = individual, NM = numeric modifier, PE = property expression, QM = quantity modifier)

ID	Pattern	Example	PA	RT	M	DE
1	Which [CE1] [OPE] [CE2]?	Which pizzas contain pork?	2	obj.		
2	How much does [CE] [DP]?	How much does Margherita Pizza weigh?	2	data.		
3	What type of [CE] is [I]?	What type of software (API, Desktop application, etc.) is it?	1			
4	Is the [CE1] [CE2]?	Is the software open-source development?	2			
5	What [CE] has the [NM] [DP]?	What pizza has the lowest price?	2	data.	num.	
6	What is the [NM] [CE1] to [OPE] [CE2]?	What is the best/fastest/most robust software to read/edit this data?	3	both	num.	
7	Where do I [OPE] [CE]?	Where do I get updates?	2	obj.		spa.
8	Which are [CE]?	Which are gluten-free bases?	1			
9	When did/was [CE] [PE]?	When was the 1.0 version released?	2	data.		tem.
10	What [CE1] do I need to [OPE] [CE2]?	What hardware do I need to run this software?	3	obj.		
11	Which [CE1] [OPE] [QM] [CE2]?	Which pizza has the most toppings?	2	obj.	quan.	
12	Do [CE1] have [QM] values of [DP]?	Do pizzas have different values of size?	2	data.	quan.	

satisfiable. Otherwise, any pizza must contain pork and the *"contains pork"* part in the CQ becomes useless.

(b) *Binary question* asks whether there is an answer satisfying the constraint. It does not have the two satisfiability presuppositions.

(c) *Counting question* asks for the number of answers satisfying the constraints. It assumes the possibility of some answer satisfying the constraint and also some answer not satisfying it. Hence it has the satisfiability presuppositions.

2. **Element Visibility:** regardless of the visibility of a modelling element, it should always occur in the ontology to make the CQ answerable. Nevertheless, an implicit element does not appear in the CQ hence its corresponding name in the ontology cannot be directly obtained. This name can be derived from related entities. For

Table 4.2 CQ Subtypes of Archetype 1 (QT = Question Type, V = Visibility, QP = Question Polarity, sel. = selection question, bin. = binary question, cout. = counting question, exp. = explicit, imp. = implicit, sub. = subject, pre. = predicate, pos. = positive, neg. = negative)

ID	Pattern	Example	QT	V	QP
1a	Which [CE1] [OPE] [CE2]?	What software can read a .cel file?	sel.	exp.	pos.
1b	Find [CE1] with [CE2]	Find pizzas with peppers and olives	sel.	imp. pre.	pos.
1c	How many [CE1] [OPE] [CE2]?	How many pizzas in the menu contain meat?	cout.	exp.	pos.
1d	Does [CE1] [OPE] [CE2]?	Does this software provide XML editing?	bin.	exp.	pos.
1e	Be there [CE1] with [CE2]?	Are there any pizzas with chocolate?	bin.	imp. pre.	pos.
1f	Who [OPE] [CE]?	Who owns the copyright?	sel.	imp. sub.	pos.
1g	Be there [CE1] [OPE]ing [CE2]?	Are there any active forums discussing its use?	bin.	exp.	pos.
1h	Which [CE1] [OPE] no [CE2]?	Which pizza contains no mushroom?	sel.	exp.	neg.

example, in "What are the export options for this software?" we can name the implicit relation *hasExportOption*. Otherwise it can be assigned by the author.

3. **Predicate Arity:** the arity of the predicate affects how it should occur in the ontology. Modern ontology languages support both unary (i.e. classes) and binary (i.e. properties) predicates. Hence their names can directly occur in the ontology. However, N-ary predicate has to be represented as a class via reification which leads to the occurrence of other implicit predicates. For example, in "What is the best software to read this data?" the predicate *read* has 3 arities, namely the *software*, the *data* and the *performance*. Hence *Reading* should occur in the ontology as a *Class* instead of a *Property*. Moreover, there should be three more implicit predicates, namely the *hasSoftware*, the *hasData* and the *hasPerformance*.
4. **Relation Type:** as the name suggests, the meta-type of a property occurring in the ontology is determined by the type of relation it represents in the CQ. In other words, if a property *P* is between two entities, then it is presupposed that *P* is an **instance** of `OWL:ObjectProperty`. If *P* is between an entity and a value, then it is presupposed that *P* is an instance of `OWL:DatatypeProperty`.
5. **Modifier:** the modifiers further impose restrictions on answers of the CQ.

   (a) **Quantity modifier** has a similar effect as *question type* on the satisfiability presupposition of certain class expressions in the ontology.

      i. If the modifier is a concrete value or range, then as for a *selection question* it presupposes that potential answers are allowed to satisfy, as well as not to satisfy, this modifier. For example, "If I have 3 ingredients, how many kinds of pizza can I make?" implies that the ontology allows pizzas with three ingredients and ones with fewer or more than three ingredients, i.e. $Pizza \sqcap = 3\ hasIngredient.Ingredient$ and $Pizza \sqcap \neg(= 3\ hasIngredient.Ingredient)$ should both be satisfiable in the ontology.

ii. If the modifier is a superlative value or value range, then the ontology should allow answers with **multiple cardinality** values on the predicate on which the modifier is imposed. For example, in "Which pizza has the most toppings?" the presupposition is that *pizzas are allowed to have different numbers of toppings* otherwise all pizzas will have exactly the same number of toppings. More formally, this means that for each number $n \geq 0$, $Pizza \sqcap \neg(= n\ hasTopping.Topping)$ should be satisfiable.

iii. If the modifier is a comparative value or value range, then the ontology should allow an answer with the required **comparative cardinality** values on the different relations being compared, as well as answers without the required comparative cardinality values. For example, "Which pizza has more meat than vegetables?" presupposes that *pizzas are allowed to have more meat than vegetables* otherwise none of the pizzas is an answer. More formally this means that for some number $n \geq 0$, $Pizza \sqcap \leq n\ has.Vegetable$ and $Pizza \sqcap \geq n + 1\ has.Meat$ should both be satisfiable. It also presupposes that *pizzas are allowed to have no more meat than vegetables* otherwise all pizzas have more meat than vegetables. More formally this means that for some number $n \geq 0$, $Pizza \sqcap \leq n\ has.Meat$ and $Pizza \sqcap \geq n\ has.Vegetable$ should both be satisfiable.

(b) *Numeric modifier* has similar presuppositions to a quantity modifier. In the concrete value or value range case and comparative value case, the CQ carries the presuppositions that the ontology should allow answers satisfying the modifier and those not satisfying the modifier. In the superlative value case, the CQ carries the presupposition that the ontology should allow multiple values on the relation on which the modifier is imposed.
Furthermore, the **range** of the property on which the modifier is imposed must be a comparable datatype, such as *integer*, or *float*, otherwise the question cannot be answered meaningfully.

6. **Domain-independent Element** in the CQ can also affect the meta-type and type of some modelling elements in the ontology. The temporal element is usually associated to some temporal datatypes. For example, "When was the 1.0 version released?" has presuppositions that the *wasReleasedOn* is a datatype property, and that the range of *wasReleasedOn* is one of the temporal datatypes such as *datatime*. It is possible to use some other datatype, such as *integer*, to denote the year of release, but this is not considered the best practice.

Table 4.3 Authoring tests

AT	Parameter	Checking
Occurrence	[E]	E in ontology vocabulary
Class satisfiability	[CE]	CE is satisfiable
Relation satisfiability	[CE1]	$CE1 \sqcap \exists P.E2$ is satisfiable, $CE1 \sqcap \neg \exists P.E2$ is satisfiable
	[P]	
	[E2]	
Meta-instance	[E1]	$E1$ has `type` $E2$
	[E2]	
Cardinality satisfiability	[CE1]	$CE1 \sqcap = nP.E2$ is satisfiable, $CE1 \sqcap \neg = nP.E2$ is satisfiable
	[n]	
	[P]	
	[E2]	
Multiple cardinality (on superlative quantity modifier)	[CE1]	$\forall n \geq 0$, $CE1 \sqcap \neg = nP.E2$ is satisfiable
	[P]	
	[E2]	
Comparative cardinality (on quantity modifier)	[CE1]	$\exists n \geq 0$, $CE1 \sqcap \leq n\ P1.E1$ and $CE1 \sqcap \geq n + 1\ P2.E2$ are satisfiable, $\exists m \geq 0$, $CE1 \sqcap \leq m\ P2.E2$ and $CE2 \sqcap \geq (m + 1)\ P1.E1$ are satisfiable
	[P1]	
	[P2]	
	[E1]	
	[E2]	
Multiple value (on superlative numeric modifier)	[CE1]	$\forall D \subseteq range(P)$, $CE1 \sqcap \neg \exists P.D$ is satisfiable
	[P]	
Range	[P]	$\top \sqsubseteq \forall P.E$
	[E]	

The *spatial element* is not necessarily representing a geographical location hence it is hard to determine the type of its corresponding element in the ontology.

Features in the CQs are related to certain categories of presuppositions. Each of these categories contains parameter(s) derived from the CQ and can be realised by some checking in the ontology. ATs formalise this idea. We summarise the ATs in Table 4.3. In this table the first column shows the ATs, the second column shows the parameters for each AT and the third column shows how each AT can be checked with ontology technologies, in which $\sqcap$ means conjunction, $\neg$ means negation, $\exists P.E$ means having P relation to some E, $= nP.E$ ($\geq nP.E$, $\leq nP.E$) means having P

relation(s) to exactly (at least, at most) n E(s), $\forall P.E$ means having P relation (if any) to only E, $\top$ means everything. We omit the formalisation of some ATs, such as those associated with comparative numeric modifiers, because such features were not observed in our collection; they can be formalised in a similar manner as the ones in the table.

As one can see, all of these ATs can be checked automatically. *Occurrence* can be checked directly against the ontology. *Meta-Instance* can be checked via RDF reasoning. All the others can be checked with ontology reasoning.

To summarise this section, we present the CQOA pipeline in Fig. 4.3 and illustrate its workflow with an example.

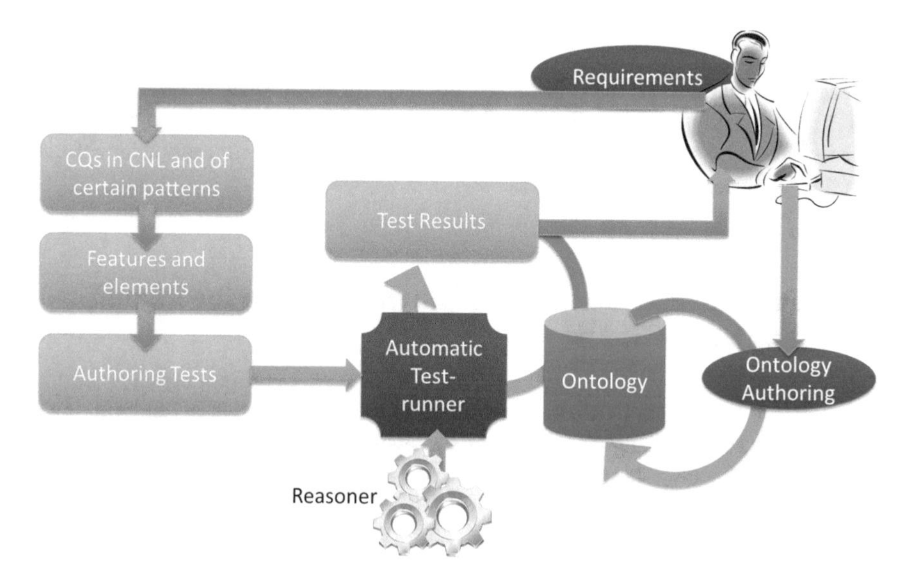

Fig. 4.3 CQOA pipeline

In an implemented system, users can use a controlled natural language (CNL) to input CQs based on the patterns identified earlier. Hence the archetype and/or the subtype of input CQs are implicitly specified by users and automatically identified by the system: For example, CQ "What is the best software to read this data?" belongs to archetype CQ pattern 7 $What$ $[CE1]$ is $[NM]$ to $[OPE]$ $[CE2]$?.

From the CQ and its pattern the system can automatically extract the features and elements of the CQ: it is a selection question ("What") containing a 3-ary (among "software", "data" and some performance) predicate ("read") with a superlative numeric modifier ("best"), which should be modelled as a class and some implicit object and datatype properties, whose names can be generated from contexts or assigned by users.

Then the system can automatically generate and parameterise the following ATs:

1. Occurrence tests of *Software*, *Data*, *Read*, *hasSoftware*, *hasPerformance* and *hasData*. The first 3 should occur as classes and the last 3 as properties. *Read* is the class representation of the "reading" predicate in the CQ;

2. Relation Satisfiability tests of (*Read*, *hasSoftware*, *Software*), (*Read*, *hasData*, *Data*) and (*Read*, *hasPerformance*, $\top$), which guarantee that the ontology allow some *Read* to be associated with *Software*, *Data* and to have performance;
3. Meta-Instance test of (*hasSoftware*, `ObjectProperty`), (*hasData*, `ObjectProperty`) and (*hasPerformance*, `DatatypeProperty`), which further specify the meta-types of the three properties;
4. Multiple Value on superlative numeric modifier test of (*Read*, *hasPerformance*), which guarantees that instances of *Read* can have *different* performance values;
5. Range test of (*hasPerformance*, *decimal* $\cup$ *float* $\cup$ *double*), which ensures that the value of *hasPerformance* must be a comparable numeric value, so that one can find the best performance;

As the pipeline shows, the procedure from CQs (in a controlled natural language) to ATs can be automated. Eventually, all these ATs can be automatically checked and the results can be provided to users. The users can then perform authoring actions with the feedback taken into account.

4.3 Semi-automated Linking of Enterprise Data for Virtual Knowledge Graphs

After having done a good knowledge modelling job, the next step in constructing a knowledge graph naturally shifts to the data level. According to the lifecycle mentioned in Sect. 3.5, this should be the data lifting step. In this section, we introduce an approach for creating data linkage that is a critical type of knowledge in knowledge graphs. Specifically, we describe Helix, a system for creating links among large-scale and heterogeneous information sources in large organisations.

Helix provides a unified view of data sources, ranging from spreadsheets and XML files with no schema, all the way to RDF graphs and relational data with well-defined schemas. Helix users explore these heterogeneous data sources through a combination of keyword searches and navigation of linked Web pages that include information about the schemas, as well as data and semantic links within and across sources. At a technical level, the section describes the research challenges involved in developing Helix, along with a set of real-world usage scenarios and the lessons learned.

4.3.1 Virtual Knowledge Graph for Knowledge Discovery

Data and/or knowledge discovery is a critical means to find relevant information for problem solving in enterprises. However, due to the Big Data challenges, it turns out to be a very challenging job especially for large organisations. Our focus in this section is on building knowledge graphs for data discovery in such scenarios. We discuss how semi-automated techniques might be brought to bear to help the problem of data discovery. We describe the data processing pipeline for supporting knowledge discovery with a special focus on interlinking techniques for constructing knowledge graphs in large organisations.

Specifically, we introduce semi-automated tagging of the data for the purposes of data discovery. With the growth of a wealth of structured semantic information about entities on the Web such as DBpedia, we show how one might construct an automatic classification system that can run over a large data pool and group the datasets into semantic types. We show how one might extend this classification system within a particular type so that one can group attributes of tables into similar types. We demonstrate how one can apply these techniques on real data pools from open city data with approximately 1000 tables. And we describe how much data still remains in if 'dark pools' after such analysis. One key point we make is that there are many possible classification techniques—we articulate one possible mechanism to automatically tag datasets, but this is doubtlessly an area for fruitful research, and we need many more techniques in this space targeting data discovery and data classification.

Once the data is thus tagged and classified, another problem that presents itself is how to use these tags effectively to help users discover relevant data. Faceted search is obviously a candidate, but that is not sufficient, as we show in our work with the city data. Links that can be established between datasets even without the use of explicit semantic tags can be very helpful in clustering subsets of the data together, as we show. It also helps users quickly understand what sorts of attributes are available to them in these datasets. As in the case of tagging, one can establish links based on multiple algorithms. Once again we describe a possible technique here based on our earlier work of large-scale schema matching which in turn is based on instance data [70], but this is also an area for further research.

Linking between attributes of datasets can also help users establish linkage points to perform fuzzy joins across the datasets. A linkage point describes a pair of schema elements that share a significant number of instance values [105]; in other words, the schema elements are good candidates for a fuzzy join between two tables. A fuzzy join is one where entity labels are matched using string similarity functions, or more sophisticated entity matching algorithms that use matching of entities across multiple columns. We describe a very simple string matching algorithm to demonstrate how such fuzzy joins may be used as part of the data discovery process. They can be invaluable in constructing appropriate datasets for future analysis, as we show in our example. Entity matching is a very well-studied problem in the database and ontology matching literature, but only when the entity type is well known and when the structure of these types is well known. In a situation when one is confronted with several thousand tables, and several thousand entity types, a different approach is required for entity matching, which is another area for fruitful research.

To summarise, the techniques that relate to the problem of data discovery include methods to (a) normalise data in different formats, (b) index structured data in tables, (c) perform semantic matching between schema elements of structured data, (d) tag data with semantic tags and (e) find linkage points in the data so that users can join between tables. In this chapter, we focus on techniques (c)–(e) and describe how one might use these components in a larger system for guided data exploration, and show use cases from actual scenarios.

4.3.2 Semantic Tagging and Data Interlinking

All input data sources in Helix are defined in the data source registry component. There are three classes of sources considered. The first class includes (semi-)structured sources with predefined schemas and query APIs, such as relational databases and triple stores. The second class is (online or local) file repositories, such as the ones published by governments (e.g. data.gov or data.gov.uk, or data sources published by U.S. National Library of Medicine), or in cloud-based file repositories (e.g. Amazon S3). Finally, the third class are those sources directly read from online Web APIs, e.g. data read using the Freebase API.

One of the goals in Helix is to process data based on explicit user needs and avoid unnecessary or expensive pre-processing given that we are dealing with large-scale enterprise data. Therefore, the data pre-processing phase comprises only three *essential* steps, all performed in a highly scalable fashion implemented in the Hadoop ecosystem: (a) schema discovery, where each input source schema is represented in a common normalised model in the form of a *local schema graph*; (b) full-text indexing, where data values and source metadata are indexed; and (c) linkage discovery that incorporates instance-based matching and clustering of the (discovered) schemas; (d) semantic tagging and schema linking, the outcome of the pre-processing phase is a semantically tagged *Global Schema Graph*; (e) linkage point discovery, the use of linkage point discovery to find possible points for fuzzy joins.

In the following, we discuss briefly steps (c)–(e).

Linkage Discovery

One key phase in data pre-processing is discovering links between different types and attributes *within* as well as *across* the schema graphs of different sources. Traditional schema-based matching is not effective in matching highly diverse and automatically constructed schemas where the labels of schema elements are not always representative of their contents, and the data come from several sources that use different models and representations. Therefore, our approach is to perform an all-to-all instance-based matching of all the attributes. Scaling the matching process for a large number of attributes and large number of instances per attribute is a major challenge. We address this problem by casting it into the problem of computing document similarity in Information Retrieval [70]. Specifically, we treat each attribute node as a document and consider the instance values for that attribute as the set of terms in the document. To scale the computation of pairwise attribute similarity, we use Locality Sensitive Hashing (LSH) techniques, as is done in computing document similarity. Briefly, we construct a fixed small number of *signature* values per attribute, based on MinHash [41] or Random Hyperplane [47], in a way that a high similarity between the set of signatures guarantees high similarity among instance values. This results in an efficient comparison of instance values between attributes. We then create small buckets of attributes, so that similar attributes are guaranteed to be in the same bucket with a high probability. This is similar to a common indexing technique used in record linkage known as *blocking* [50]. Our experiments on large data sources show that our approach is very effective in reducing the number of pairwise attribute comparisons required for an all-to-all attribute matching [70].

In our evaluation, we found that the precision and recall of linkages between attributes with textual values is very good [70]. However, linkages between attributes with numeric or date/time values tend to have little semantic value, even when the similarity of their instances is high. Currently, we optionally filter attributes with these data types. We are investigating the scalability of *constraint-based instance matching* [197] for discovering linkages between such attributes.

The attribute-level linkages found within and across data sources are used not only for guided navigation of the sources, but also to find type-level linkages and grouping (clustering) of types. In more detail, type clustering is performed to group types that have the same or highly similar attribute sets. For example, all 'address' types of an XML source might create a single cluster, in spite of these addresses appearing at different levels and under different elements of the tree. Type-level linkages induce a similarity graph, where each node represents a type and the weight of an edge connecting two types reflects their similarity. This similarity is the average of (a) the instance-based similarity between the attributes of the two types; and (b) the Jaccard similarity between the sets of attribute labels of the types. An unconstrained graph clustering algorithm [104] is then used to find clusters of types in the similarity graph.

Semantic Tagging and Schema Linking

The schema graphs of all the input sources along with discovered attribute and type linkages are all used to build the *Global Schema Graph*. This graph provides a unified view over the input sources, enables navigation and allows the discovery of related attributes and types through schema and similarity-based linkages. Figure 4.4 shows a portion of a global schema graph constructed for one of our use cases. In this example, a dataset on national heritage sites in the city of Dublin is linked to a dataset in the same source containing school locations, based on the similarity of the address/location attributes in the two datasets. The dataset is also linked to a semantic type in a Web knowledge base that contains information on architectural buildings, which itself is linked to another knowledge base containing information about public locations (*Place* type in an ontology). These links implicitly show that these datasets contain information about locations, and that there is potentially a connection between school locations and national heritage sites in the city of Dublin, one of the many exploration capabilities of Helix. In the figure, we distinguish two

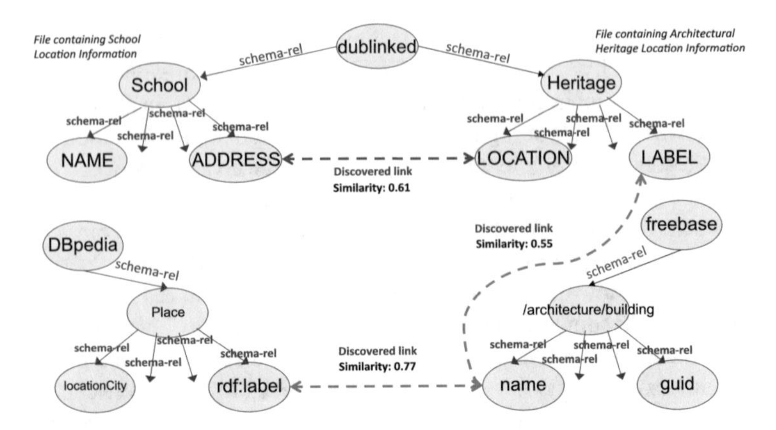

Fig. 4.4 Sample schema graph

sorts of links, namely *explicit* links (drawn in solid black lines) that are inferred by looking at individual sources through schema discovery, and *discovered* links (drawn in dashed blue lines) that require additional logic and consider multiple sources (see Sect. 4.3.2). For discovered links, we add annotations to capture their computed similarity, as well as the method by which the link was discovered (e.g. MinHash, user generated, etc.).

The global schema graph is a key structure in Helix because it governs and guides user interactions. What is less obvious though is that there are technical challenges in terms of managing the graph itself. Helix is geared towards Big Data scenarios, and as more and more sources are incorporated into the system, the global schema graph very quickly becomes quite large. As the system continuously queries, updates and augments the graph, it is important that all these operations be performed efficiently; otherwise the global schema graph ends up being a bottleneck to the system's performance. To address these challenges, we store the global schema graph in our own graph store, called DB2RDF, which has been proven to outperform competing graph stores in a variety of query workloads using both synthetic and real data [36]. Our graph store supports the SPARQL 1.0 graph query language [191] and interactions with the global query graph are automatically and internally translated to SPARQL queries.

Linkage Point Discovery

Hassanzadeh et al. [105] describe a set of algorithms called SMaSh-S, SMaSh-R and SMaSh-X, respectively, to find linkage points in the data. In our case study, we used the SMaSh-R algorithm to perform the linkage point discovery. At its core, the SMaSh-R algorithm takes a sample of instance values for a particular source schema element t_1, and tries to find a corresponding set of instance values in schema element t_2, using indexing techniques to find possible matches. In indexing the instance values, each instance value is treated as a document, which means that the degree to which a target instance value matches the source instance value in the text index is governed by the number of matching tokens (along with factors such as TF-IDF). In Hassanzadeh et al., each schema element is compared with every other schema element because of the constrained number of types they matched. In our scenario, such an n^2 comparison will not scale. Therefore, we pruned the set of pairs to be compared based on the results of the schema matching algorithm we described earlier. Once each linkage point is computed for all possible schema matchings, we filter the linkage points as specified in Hassanzadeh et al. such that linkage points with low cardinality, low coverage or low strength could be eliminated. These linkage points may be seen as extending the global schema graph with a different set of links. The links now refer to points that can be used to join tables together based on common instance values.

4.3.3 Usage Scenarios

The design and implementation of the Helix system has gone through extensive evaluation using several usage scenarios in different domains. The majority of the usage scenarios are inspired by our interactions with customers, in trying to understand their needs in data exploration and help them with the first steps of their data analytics tasks. In this subsection, we describe three of such usage scenarios and some of our key observations and lessons learned. We first describe details of two usage scenarios using data published by the cities of Ireland, New York and San Francisco. Extracting relevant information from online public data repositories such as those published by government agencies is a frequent request within enterprises. We then describe a customer relationship management (CRM) use case as an example enterprise data exploration scenario.

Note that our goal here is not to perform a scientific study of the effectiveness of the algorithms implemented in the system (such as the study we had performed on the accuracy of attribute-level linkages [70] and linkage point discovery [105]). Nor do we intend to evaluate the effectiveness of our user interface through a large-scale user study, which is a topic of future work and beyond the scope of this section.

Table 4.4 provides a summary of the source characteristics in the three scenarios, and Table 4.5 provides the total number of links found across these sources. Each source is in itself composed of multiple datasets. We therefore provide a summary of the number of links between data sources, as well as the summary of links within a single data source (e.g. a single data source like the NYC open data is composed of several hundred files). The number of links is provided to demonstrate that the system computes a large number of them. It is not our intent here to characterise them by the standard metrics of precision and recall (cf. [70]). As the links are used primarily within the context of a rather focused search, we illustrate in the use cases below how sample links may help data discovery and analysis.

Table 4.4 Summary of data sources

Data source	Types	Instances	Tables/files
Bug reports	201	7M	1
Bug fixes	95	121M	7
Freebase	1,069	44M	NA
DBpedia	155	2M	NA
Dublinked	1,569	22M	485
NYC/SF	971	30M	971

Table 4.5 Links across data sources' types

Data Src/Data Src	#Links	Data Src/Data Src	#Links
Bug fixes/Bug fixes	1,510	Bug reports/Freebase	298
Bug fixes/Bug reports	1,209	Bug reports/Bug reports	316
Bug fixes/DBpedia	25	Dublinked/Dublinked	288,045
Bug fixes/Freebase	1,216	Dublinked/DBpedia	225
Bug reports/DBpedia	4	Dublinked/Freebase	2,351
SF/SF	329,126	NYC/NYC	5,843,896
SF/NYC	919,569	NYC/DBPedia	3,868
SF/DBPedia	867		

Scenario 1: Dublinked—Open Data in Dublin

The city of Dublin has a set of data from different government agencies that is published in a number of different file formats (see: http://dublinked.ie/). At the time of this writing, Helix could access 203 collections. Each collection consists of multiple files, resulting in 501 files with supported formats that broke down into 206 XLS, 148 CSV, 90 DBF and 57 XML files. Helix indexed and pre-processed 485 files, but 16 files could not be indexed due to parsing errors. Our main use case here is data integration across the different agency data, but we also decided to connect the Dublinked data to Freebase and DBpedia, to determine if we could use the latter two sources as some form of generic knowledge. For the pre-processing step, we processed DBpedia and Freebase as RDF dumps.

The value of integrating information across files and across government agencies is obvious, but we illustrate here a few examples, based on links discovered in our pre-processing step, in Table 4.6. Here are some examples of questions that a city official can now construct queries for, based on the Helix discovered linkages in the data shown in the table:

1. Find schools that are polling stations, so that the city can prepare for extra traffic at schools during voting periods.
2. Find disabled parking stations that will be affected by pending gully repairs, to ensure accessibility will be maintained in a specific region of the city.

Table 4.6 Sample links for the Dublinked scenario

Property pairs	Score
`xml://School-Enrollment/Short-Name` → `xml://Polling-Stations-table/Name`	0.82
`csv://DisabledParkingBays/Street` → `csv://GullyRepairsPending/col2`	0.68
`xls://CanRecycling/col0` → `xls://GlassRecycling/col0`	0.71
`csv://PostersPermissionsOnPoles/Org` → `csv://CandidatesforElection2004/col2`	0.54
`csv://CandidatesforElection2004/col1` → `csv:/CandidatesforLocalElection2009/col5`	0.97
`csv://PlayingPitches/FACILITY-NAME` → `csv://PlayAreas/Name`	0.40
`csv://FingalNIAHSurvey/NAME` → `http://rdf.freebase.com/architecture/structure/name`	0.56
`dbf://Nature-Development-Areas/NAME` → `http://rdf.freebase.com/sports/golf-course/name`	0.55
`csv://ProtectedStructures/StructureName` → `http://dbpedia.org/HistoricPlace/label`	0.42

3. Find recycling stations that handle both cans and glass to route waste materials to the right stations.
4. Find organisations which have the most number of permissions to put posters on poles, to assess organisations with maximal reach to citizens.

In general, links alert users to the possibility of related data that could be pooled before any analytics is performed. For instance, any analytics on play areas would likely need to include the data in PlayAreas file as well as in the Play pitches file. Similarly, time series analysis of election data would likely include the 2004 file as well as the 2009 file. Finally, links to external datasets can easily imbue the data with broader semantics. As examples, the Name column in the FingalNIAHSurvey file refers to architectural structures, but another column also called 'Name' in the Nature-Development-Areas file is really about golf courses or play areas. Similarly, the StructuredName column in the ProtectedStructures file is about historic structures.

Table 4.7 shows two sample type clusters that Helix discovered in the Dublinked data. Recall that type clusters are based on the similarity of the schema elements in the type, as well as the instance similarity of each of those elements. The first cluster (files starting with `SchoolEnrollment*`) groups data by year despite changes in the data format. The second cluster (files starting with `GullyCleaningDaily 2004*`) discovered by Helix groups data by area, as is apparent from the titles of the

Table 4.7 Type clusters for the Dublinked scenario

Type clusters
xml://SchoolEnrollment20092010-1304
csv://SchoolEnrollment20102011-2139
xml://SchoolEnrollment20102011-2146
csv://SchoolEnrollment20082009-1301
xml://Schoolenrollment20082009-1303
csv://GullyCleaningDaily2004-11CENTRALAREA-1517
csv://GullyCleaningDaily2004-11NORTHCENTRALAREA-1518
csv://GullyCleaningDaily2004-11NORTHWESTAREA-1518
csv://GullyCleaningDaily2004-11SOUTHEASTAREA-1519

files. Following our design goals in Helix, the system itself is not trying to interpret the semantics of each discovered cluster. It will provide a tool for a knowledge worker to specify datasets for meta-analysis.

Obviously the Helix system is incapable of interpreting the semantics of each discovered data cluster, but once again, as a tool for a knowledge worker, the discovery of these clusters is invaluable in specifying datasets for meta-analysis.

Scenario 2: Open Data in NYC/SF

The cities of New York (NY) and San Francisco (SF) publish their data through the Socrata platform. NY and SF are two of the increasingly growing cities in the list of US cities that have set in place strong policies to facilitate data transparency and accessibility, and which require their agencies to constantly update and publish their datasets through the cities' platforms. The New York City (NYC) and SF Socrata platforms offer basic access facilities to these datasets, allowing users to search the data by keyword and explore by general topics. Users can then preview or download the results. Our main use case for these datasets is data integration and exploration, which we will now describe while highlighting the features of Helix that make it possible.

Consider the case of a knowledge worker looking for data on the quality of service in hospitals in the NYC. In order to do this she may search for the keyword *Hospital* in the NYC's open data Web site. However, using this method would require considerable amount of time to filter the results, because many of the results are datasets that contain the word *hospital* without actually containing any hospital data. For example, one among such non-related results includes the *School Attendance and Enrollment by District 2010–11* table. Likewise, browsing the topic *Health* (one of the 11 categories of the data) returns too many results to be inspected.

Instead, she can use Helix to link the NYC dataset to DBpedia and browse the datasets through a larger, and better organised list of topics, i.e. the classes in the hierarchies of DBpedia. To do this, we downloaded the NYC and SF data by crawling their Socrata portals. We fetched a total of 971 files (CSV and XML). Helix created a total of 4,012,385 similarity links, which we summarise in Table 4.5. Using these links we can compute the similarity of the tables in the datasets with the classes in DBpedia and use these to recommend to the user tables related to the DBpedia topic of interest, as seen in Fig. 4.5.

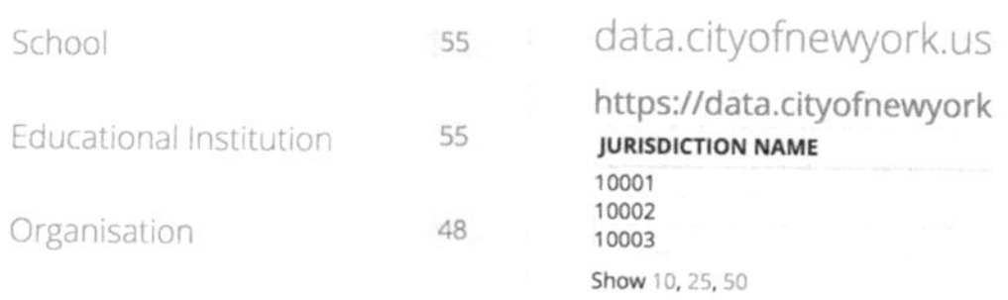

Fig. 4.5 DBpedia facets for the NYC/SF data

The knowledge worker can now select *Hospital* to view all the tables associated to the DBpedia class Hospital. The result is a list of five tables (see Fig. 4.6), one of them clearly containing information about the quality of service.

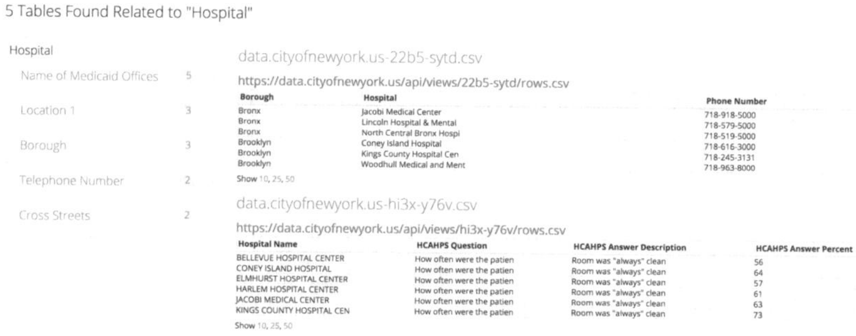

Borough	Hospital	Phone Number
Bronx	Jacobi Medical Center	718-918-5000
Bronx	Lincoln Hospital & Mental	718-579-5000
Bronx	North Central Bronx Hospi	718-519-5000
Brooklyn	Coney Island Hospital	718-616-3000
Brooklyn	Kings County Hospital Cen	718-245-3131
Brooklyn	Woodhull Medical and Ment	718-963-8000

Hospital Name	HCAHPS Question	HCAHPS Answer Description	HCAHPS Answer Percent
BELLEVUE HOSPITAL CENTER	How often were the patien	Room was "always" clean	56
CONEY ISLAND HOSPITAL	How often were the patien	Room was "always" clean	64
ELMHURST HOSPITAL CENTER	How often were the patien	Room was "always" clean	57
HARLEM HOSPITAL CENTER	How often were the patien	Room was "always" clean	61
JACOBI MEDICAL CENTER	How often were the patien	Room was "always" clean	63
KINGS COUNTY HOSPITAL CEN	How often were the patien	Room was "always" clean	73

Fig. 4.6 DBpedia Hospital facet for the NYC/SF data

If the user selects this table, Helix will now recommend tables that are similar to the current table. These tables may be used to create UNION views (explained in the previous sections) or they can be *joined* with the current table. This allows the user to take advantage of the complementary data available in other tables. For example, here Helix presents an additional *linkable* table that contains general information about hospitals, e.g. telephone and the borough where the hospital is located. This is information that is not available in the *hospital satisfaction table* and that is pertinent to the hospitals available in such a table.

By clicking on "explore linkage" for this table, the user is presented with a view created from the join of the two tables. Helix presents the user with all possible "linkage points", i.e. pairs of columns that can be joined. In this case, there is only one choice and it is already selected, the join of the `Hospital` and `Hospital Name` columns. Note that the join is in fact a *fuzzy join* in the sense that the instances do not match perfectly, e.g. they may have slight wording variations or case usage, e.g. Metropolitan Hospital versus METROPOLITAN HOSPITAL CENTER. The resulting view can be saved for later use or exported as a CSV file.

Scenario 3: Customer Relation Management in Large Organisations

In most enterprises, maintaining a consistent view to customers is key to customer relationship management (CRM). This task is made difficult by the fact that the notion of a customer frequently changes with business conditions. For instance, if an enterprise has a customer *"IBM"* and also a customer *"SoftLayer"*, they are distinct entities up until the point that one acquires the other. After the acquisition, the two resolve to the same entity. The process of keeping these entities resolved and up to date in the real world is often a laborious manual process, which involves looking up mergers and acquisitions on sites like Wikipedia and then creating scripts to unify the companies involved. Our second scenario targets this use case. The real-world sources involved are (a) a relational database with a single table that tracks defects against products, (b) a relational database that tracks fixes for the defects in the defect tracking database (with seven tables—one per product) and (c) an RDF version of Wikipedia, from Freebase/DBpedia.

The query that the knowledge worker is interested in is a picture of the number of defects fixed for each customer (where each customer is grouped or resolved by merger and acquisition data). We highlight the features of Helix that help the user build this query. We illustrate what steps a user would take in Helix if her intent is to build a table of customer records of bugs and their corresponding fixes, accounting for the latest mergers and acquisitions. Note that because much of this data is proprietary, we do not display confidential results.

Step 1 The user issues a keyword search on a customer name, such as 'IBM', to see what they can find. The hits returned include records in the bug database, as well as nodes in the Freebase/DBpedia RDF graph which match IBM (e.g. IBM, IBM AIX, etc.). The user then clicks a particular hit in the bug report database to explore the record in the context of the original table/graph. The user sees the larger context for the table (other records in the column that contains IBM, and other columns in the table that are related to the `CUST-NAME` column within the same table). More importantly, the user finds other properties that also contain similar data (as an example, see Table 4.8 that shows some real links found by Helix). If the user browses the `CUST-NAME`

Table 4.8 Sample links for the CRM scenario

Property pairs	Score
rdb://Bug-Reports/CUST-NAME → rdb://Bug-Fixes/Product1/CUSTOMER-NAME	0.75
rdb://Bug-Reports/CUST-NAME → rdb://Bug-Fixes/Product2/CUSTOMER-NAME	0.74
rdb://Bug-Reports/CUST-NAME → http://rdf.freebase.com/business-operation/name	0.47
rdb://Bug-reports/CUST-NAME → http://dbpedia.org/Company/label	0.28

column in the bug report database, Helix recommends the `CUSTOMER` column in the bug fixes database, and the `/business/organization` type in the RDF Freebase graph based on the links.

Step 2 The next step is the creation by the user of multiple virtual views that are saved to iteratively reach the desired result (see Fig. 4.7). Step 2 is a direct outcome of the data exploration conducted by the user in Step 1, where the user finds relevant data, and now wants to subset it for their task. For this example, we assume the user creates three virtual views. The first view contains a subset of bug reporting data with the columns `CUSTOMER` and `BUG NUMBER`, the second contains a subset of the bug fixes data with the columns `BUG NO`, `FIX NO`, and the third contains a subset of Freebase data, with `/business/employer`, and its relationship to its acquisitions through the `/organization/companiesAcquired` attribute.

Step 3 This step involves using semantic joins to build more complex views customised for the user's task. Here, the user likely joins Views 1 and 2 on `BUG NUMBER` and `BUG NO` to create View 4 of bugs that were fixed for different customers. Then, the user joins Views 3 and 4 on `CUSTOMER` and `/organization/companiesAcquired` to create View 5 of bugs and fixes by customer, where the customer record also reflects any companies acquired by customers in the bug report/fixing sources. At this point, the user could union View 4 with View 5 to find a count of bugs and fixes delivered to a customer and any of its acquisitions. Figure 4.7 shows all the steps in the process. In the figure, notice that a bug like 210, which normally would only be attributed to customer *"SoftLayer"*, is now also counted as part of the bugs for customer *"IBM"* since the latter acquired the former. Knowledge of these acquisitions can be used to further *refine* the result by, say, removing all *"SoftLayer"* entries because they are already considered as part of *"IBM"*.

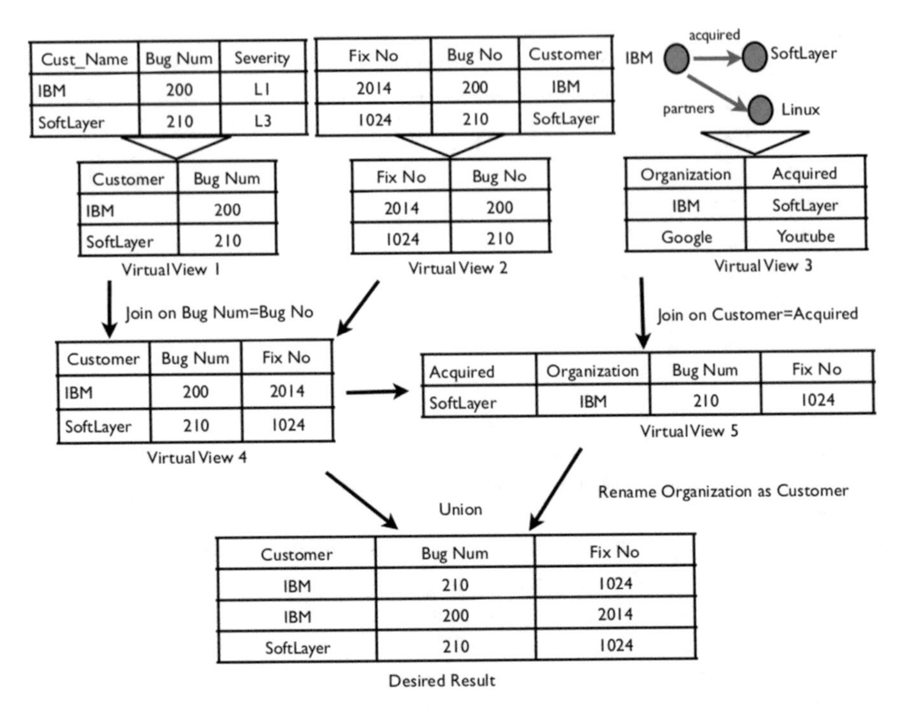

Fig. 4.7 Steps in the CRM scenario

4.3.4 Conclusion

In this section, we described Helix, a system that allows knowledge workers and data scientists to *explore* a large number of data sources using a unified intuitive user interface. Users can find portions of the data that are of interest to them using simple keyword search, navigate to other relevant portions of the data and iteratively build customised views over one or more data sources. These features rely on highly scalable schema and linkage discovery performed as a pre-processing step, combined with online (and in part social) guidance on linkage and navigation. We demonstrated capabilities of our system through a number of usage scenarios.

Chapter 5
Construction of Enterprise Knowledge Graphs (II)*

Panos Alexopoulos, Yuting Zhao, Jeff Z. Pan and Man Zhu

In this chapter, we continue with the *Acquisition and Integration Layer* of Chap. 3's reference architecture, focusing on knowledge graph construction techniques. Nevertheless, we shift from semi-automated approaches to automated approaches of knowledge graph construction by describing two additional frameworks, one for entity/scope resolution of textual data (Sect. 5.1) and one for the learning of ontological schemas from data (Sect. 5.2).

5.1 Scenario-Driven Named Entity and Thematic Scope Resolution of Unstructured Data*

As already suggested in Sect. 3.2, the task of *Named Entity Resolution* involves detecting mentions of named entities (e.g. people, organisations or locations) within texts and mapping them to their corresponding entities in a given knowledge source. The typical problem in this task is ambiguity, i.e. the situation that arises when

---

P. Alexopoulos (✉)
Expert System, Prof. Waksman 10, 28036 Madrid, Spain
e-mail: palexopoulos@expertsystem.com

Y. Zhao
IBM Italia, Circonvallazione Idroscalo, 20090 Milan, Italy
e-mail: yuting.zhao@it.ibm.com

J.Z. Pan
University of Aberdeen, King's College, Aberdeen AB24 3UE, UK
e-mail: jeff.z.pan@abdn.ac.uk

M. Zhu
Southeast University, 2 Sipailou, Xuanwu, Nanjing 210018, China
e-mail: mzhu@seu.edu.cn

J.Z. Pan et al. (eds.), *Exploiting Linked Data and Knowledge Graphs in Large Organizations*, DOI 10.1007/978-3-319-45654-6_5

a term may refer to multiple different entities. For example, "Tripoli" may refer, among others, to the capital of Libya or to the city of Tripoli in Greece. On the other hand, the **Thematic Scope** of a document can be defined as the set of semantic entities the document actually talks about. For example, the scope of a film review is typically the film the review is about while a biographical note's scope includes the person whose life is described.

In this section, we describe Knowledge Tagger (KT), a framework that performs Named Entity and Thematic Scope Resolution in texts using relevant domain ontologies and semantic data as background knowledge. Its distinguishing characteristic is its disambiguation-related customisation capabilities as it allows users to define and apply custom disambiguation evidence models, based on their knowledge about the domain(s) and expected content of the texts to be analysed.

5.1.1 Framework Description

Knowledge Tagger facilitates Named Entity and Thematic Scope Resolution in application scenarios where:

- The documents' domain(s) and content nature are a priori known or can be predicted.
- Comprehensive ontologies covering these domain(s) are available (either purposely built or from existing sources such as Linked Data).

By *content nature*, we practically mean the types of semantic entities and relations that are expected to be found in the documents. For example, in film reviews one can expect to find films along with directors and actors that have directed them or played in them respectively. Similarly, in texts describing historical events one will probably find, among others, military conflicts, locations where these conflicts took place and people and groups that participated in them. Documents with known content nature, like the above, can be found in many application scenarios where content is specialised and focused (e.g. reviews, scientific publications, textbooks, reports, etc.).

Given such scenarios, KT targets the two tasks of Named Entity and Thematic Scope Resolution based on a common intuition: a given ontological entity is more likely to represent the meaning of an ambiguous term or fall within the thematic scope of a text when the latter contains several additional entities that are ontologically related to it. These related entities can be seen as **evidence** whose quantitative and qualitative characteristics can be used to determine the most probable meaning of the term.

To see why this assumption makes sense, assume a historical text containing the term "Tripoli". If this term is collocated with terms like *"Siege of Tripolitsa"* and *"Theodoros Kolokotronis"* (the commander of the Greeks in this siege) then it is fair to assume that this term refers to the city of Tripoli in Greece rather than the capital of Libya. Also, in a cinema-related text like *"Annie Hall is a much better movie than Deconstructing Harry, mainly because Alvy Singer is such a well formed character and Diane Keaton gives the performance of her life"*, the evidence provided by the entities *"Alvy Singer"* (a character in the movie Annie Hall) and *"Diane Keaton"* (an actress in the movie Annie Hall) indicates that *"Annie Hall"* is more likely to be the movie the text is about rather than, e.g. *"Deconstructing Harry"*.

Nevertheless, which entities are (and to what extent they are) potential evidences in a given application scenario depends on the domain and expected content of the texts that are to be analysed. For example, in the case of historical texts we expect to use as evidence historical events and persons who have participated in them. For that reason, our approach is based on the priori determination and acquisition of the optimal evidential knowledge for the scenario in hand. This knowledge is expected to be available in the form of an ontological knowledge graph and it is used within the framework to perform geographical entity and scope resolution. The framework components that enable this feature are the following:

- An **Entity and Thematic Scope Resolution Evidence Model** that contains, for a given scenario, the entities that may serve as resolution evidence for the scenario's target entities (i.e. entities we want to disambiguate and entities that possibly denote the text's scope). Each pair of a target entity and an evidential one is accompanied by a degree that quantifies the latter's evidential power for the given target entity.
- An **Evidence Model Construction Process** that builds, in a semi-automatic manner, an entity and thematic scope resolution model for a given scenario.
- A **Entity Resolution Process** that uses the evidence model to detect and extract from a given list of text terms that refer to the scenario's target entities. Each term is linked to one or more possible entity URIs along with a confidence score calculated for each of them. The entity with the highest confidence should be the one the term actually refers to.
- An **Thematic Resolution Process** that uses the evidence model to determine, for a given text, the entities that potentially comprise the text's thematic scope. A confidence score for each entity is used to denote the most probable ones.

In the following paragraphs, we elaborate on each of the above components.

Evidence Model and Its Construction

For the purposes of this section, we define an ontology as a tuple $O = \{C, R, I, i_C, i_R\}$ where

- C is a set of concepts.
- I is a set of instances.
- R is a set of binary relations that may link pairs of concept instances.
- i_C is a concept instantiation function $C \rightarrow I$.
- i_R is a relation instantiation function $R \rightarrow I \times I$.

Given an ontology, the **Entity and Thematic Scope Resolution Evidence Model** defines which ontological instances and to what extent should be used as evidence towards (i) the correct meaning interpretation of an entity term to be found within the text and (ii) the correct thematic scope resolution of the whole text. More formally, a thematic scope evidence model consists of two functions:

- An **entity disambiguation evidence function** $edef : I \times I \rightarrow [0, 1]$. If $i_1, i_2 \in I$ then $edef(i_1, i_2)$ is the degree to which the existence, within the text, of i_2 should be considered an indication that i_1 is the correct meaning of any text term that has i_1 within its possible interpretations.
- A **thematic scope evidence function** $tsef : I \times I \rightarrow [0, 1]$. If $i_1, i_2 \in I$ then $tsef(i_1, i_2)$ is the degree to which the existence, within the text, of i_2 should be considered an indication that i_1 belongs to the thematic scope of the text.

These two functions are expected to be constructed prior to the execution of the resolution process through a semi-automatic process. The manual part of this process is executed first and includes the following steps:

1. Based on the resolution scenario, we determine the ontology concepts whose instances are expected to comprise the scope of the texts (e.g. the concept **"Film"** in the film review scenario or the concept **"Location"** in the historical text scenario).
2. For each of these concepts, we determine the related to them concepts whose instances may serve as a thematic scope evidence (e.g. **"Actor"** and **"Director"** in the film review scenario or **"Military Conflict"** and **"Military Person"** in the historical text one).
3. For each evidence concept we determine, in a similar way, the related to them concepts whose instances may serve as a disambiguation evidence.

The result of this analysis should be two concept mapping functions:

- A thematic scope evidence mapping function $tsev_C : C \rightarrow C \times R^n$ which given a scope concept $c_s \in C$ returns the concepts which may act as a thematic scope evidence for it along with the ontological relation (or chain of relations) that links this concept to the target one.
- A disambiguation evidence function $dev_C : C \rightarrow C \times R^n$ which given an evidence concept $c_e \in C$ returns the concepts which may act as a disambiguation evidence for it along with the ontological relation (or chain of relations) that link this concept to the target one.

Table 5.1 contains an example of a thematic scope evidence mapping for the military conflict texts scenario where, for instance, military conflicts provide scope-related evidence for the locations they have taken place and military persons provide evidence for locations they have fought in. The latter mapping, shown in the third row of the table, is facilitated by the chain of two relations: (i) the inverse of the relation **dbpprop:commander** that relates persons with battles they have commanded and (ii) the relation **dbpprop:place** that relates battles to their locations). In a similar

Table 5.1 Sample evidence concept mapping for military conflict texts (from DBPedia)

Scope concept	Evidence concept	Relation(s) linking evidence to target
dbpedia-owl:PopulatedPlace	dbpedia-owl:MilitaryConflict	dbpprop:place
dbpedia-owl:PopulatedPlace	dbpedia-owl:MilitaryConflict	dbpprop:place, dbpedia-owl:isPartOf
dbpedia-owl:PopulatedPlace	dbpedia-owl:MilitaryPerson	is dbpprop:commander of, dbpprop:place
dbpedia-owl:PopulatedPlace	dbpedia-owl:PopulatedPlace	dbpedia-owl:isPartOf

way, one may define a disambiguation evidence mapping for the same scenario by, for example, considering military conflicts mentioned in the text as an evidence for the disambiguation of military persons.

Using the two evidence concept mapping functions $tsev_C$ and dev_C, we can then automatically derive the corresponding evidence model functions *tsef* and *edef* as follows. Given a scope concept $c_s \in C$ and a scope evidence concept $c_{se} \in C$ then for each instance $i_s \in i_C(c_s)$ and $i_{se} \in i_C(c_{se})$ that are related to each other through the composition of relations $\{r_1, r_2, ..., r_n\} \in tsev_C(c_s)$ we derive the set of instances $I_s \subseteq I$ which are also related to i_{se} through $\{r_1, r_2, ..., r_n\} \in tsev_C(c_s)$. Then the value of *tsef* for this pair of instances is computed as follows:

$$tsef(i_s, i_{se}) = \frac{1}{|I_s|} \tag{5.1}$$

The intuition behind this formula is that the scope evidential power of a given entity is inversely proportional to the number of different target entities it provides evidence for. If, for example, a given actor has played in many different films, then its scope evidential power for this film is low.

Similarly, given a scope evidence concept $c_{se} \in C$ and a disambiguation evidence concept $c_{de} \in C$ then for each instance $i_{se} \in i_C(c_{se})$ and $i_{de} \in i_C(c_{de})$ that are related to each other through the composition of relations $\{r_1, r_2, ..., r_n\} \in dev_C(c_{se})$ we derive the set of instances $I_{se} \subseteq I$ which share common names with i_{se} and are also related to i_{de} through $\{r_1, r_2, ..., r_n\} \in dev_C(c_{se})$. Then the value of *edef* for this pair of instances is computed as follows:

$$edef(i_{se}, i_{de}) = \frac{1}{|I_{se}|} \tag{5.2}$$

Again, the intuition here is that the disambiguation evidential power of a given entity is inversely proportional to the number of different target entities it provides evidence for. If, for example, a given military person has fought in many different locations with the same name, then its evidential power for this name is low.

Entity Reference Resolution Process

The entity reference resolution process for a given text document and a disambiguation evidence model starts by extracting from the text the set of terms T that match to some instance belonging to a target or an evidence concept. Along with that we derive a term-meaning mapping function $m : T \rightarrow I$ that returns for a given term $t \in T$ the instances it may refer to. We also consider I_{text} to be the superset of these instances.

Then we consider the set of potential target instances found within the $I^t_{text} \subseteq I_{text}$ and for each $i_t \in I^t_{text}$ we derive all the instances i_e from I_{text} for which $edef(i_t, i_e) > 0$. Subsequently, by combining the evidence model $edef$ with the term-meaning function m we are able to derive an entity-term disambiguation support function $sup_d : I^t_{text} \times T \rightarrow [0, 1]$ that returns for a target entity $i_t \in I^t_{text}$ and a term $t \in T$ the degree to which t supports i_t:

$$sup_d(i_t, t) = \frac{1}{|m(t)|} \sum_{i_e \in m(t)} edef(i_t, i_e) \tag{5.3}$$

Using this function we are able to calculate for a given term in the text the confidence that it refers to the entity $i_t \in m(t)$ as follows:

$$conf_d(i_t) = \frac{\sum_{t \in T} K(i_t, t)}{\sum_{i'_t \in m(t)} \sum_{t \in T} K(i'_t, t)} * \sum_{t \in T} sup_d(i_t, t) \tag{5.4}$$

where $K(i_t, t) = 1$ if $sup(i_t, t) > 0$ and 0 otherwise. In other words, the overall support score for a given candidate target entity is equal to the sum of the entity's partial supports (i.e. function sup) weighted by the relative number of terms that support it. It should be noted that in the above process we adopt the one referent per discourse approach which assumes one and only one meaning for a term in a discourse.

Thematic Scope Resolution Process

To perform thematic scope resolution we first need to determine which are the candidate scope entities. To do that we consider the extracted text terms and map them, through function (5.4), to the scope evidence entities they most likely refer to, namely the entities with the highest confidence score. The result of this mapping is a new term-meaning function $m'(t) : T \rightarrow I$ that returns for a given term the single highest confidence entity it refers to. Given that, the candidate scope entities are those for which we have found corresponding evidence terms within the text, namely the set I_{cand} where $\forall\, i_{cand} \in I_{cand}\ \exists\, t \in T$ such that $tsef(i_{cand}, i_{se}) > 0, i_{se} \in m'(t)$.

Subsequently, we derive an entity-term scope support function $sup_s : I^{cand} \times T \rightarrow [0, 1]$ that returns for a candidate scope entity $i_{cand} \in I_{cand}$ and a term $t \in T$ the degree to which t supports the "candidacy" of i_{cand}:

$$sup_s(i_{cand}, t) = tsef(i_{cand}, i_{se}) \cdot conf_d(t, i_{se}), i_{se} \in m'(t) \tag{5.5}$$

Finally we compute, in the same way as in equation (5.6), for each given candidate scope entity $i_{cand} \in I_{cand}$ the confidence that it actually belongs to the thematic scope of the text as follows:

$$conf_s(i_{cand}) = \frac{\sum_{t \in T} K(i_{cand}, t)}{\sum_{i' \in I_{cand}} \sum_{t \in T} K(i', t)} \cdot \sum_{t \in T} sup_s(i_{cand}, t) \tag{5.6}$$

where $K(i_{cand}, t) = 1$ if $sup_s(i_{cand}, t) > 0$ and 0 otherwise.

5.1.2 Framework Application Evaluation

To assess the applicability and effectiveness of Knowledge Tagger in linking texts to knowledge graphs, we have applied it in different scenarios and domains, described below.

Scenario 1: Resolving Teams and Players in Football-Related Texts

In this scenario, we had to semantically annotate a set of textual descriptions of football match highlights like the following: *"It is the 70th minute of the game and after a magnificent pass by Pedro, Messi managed to beat Claudio Bravo. Barcelona now leads 1-0 against Real."* These descriptions were used as metadata of videos showing these highlights and our goal was to determine, in an unambiguous way, who were the participants (players, coaches and teams) in each video.

To achieve this goal, we applied our framework and built a disambiguation evidence model, based on DBpedia, that had as a disambiguation evidence mapping function that of Table 5.2. This function was subsequently used to automatically calculate the function *edef* for all pairs of target and evidence entities. Table 5.3 shows a small sample of these pairs where, for example, Getafe acts as evidence for the disambiguation of Pedro León because the latter is a current player of this football team. Its evidential power, however, for that player is 0.5, because in the same team there is yet another player with the same name (i.e. Pedro Ríos Maestre).

Using this model, we applied our disambiguation process in 50 of the above texts, all containing ambiguous entity references. Table 5.4 shows the results achieved by our approach as well as by DBPedia Spotlight and AIDA.

Table 5.2 Sample disambiguation evidence concept mapping for football match descriptions

Target concept	Evidence concept	Relation(s) linking evidence to target
dbpedia-owl:SoccerPlayer	dbpedia-owl:SoccerClub	is dbpprop:currentclub of
dbpedia-owl:SoccerPlayer	dbpedia-owl:SoccerPlayer	dbpprop:currentclub, is dbpprop:currentclub of
dbpedia-owl:SoccerClub	dbpedia-owl:SoccerPlayer	dbpprop:currentclub
dbpedia-owl:SoccerClub	dbpedia-owl:SoccerManager	dbpedia-owl:managerClub
dbpedia-owl:SoccerManager	dbpedia-owl:SoccerClub	is dbpedia-owl:managerClub of

Table 5.3 Examples of target-evidential entity pairs for the football scenario

Target entity	Evidential entity	dem
dbpedia:Real_Sociedad	dbpedia:Claudio_Bravo_(footballer)	1.0
dbpedia:Pedro_Rodriguez_Ledesma	dbpedia:FC_Barcelona	1.0
dbpedia:Pedro_Leon	dbpedia:Getafe_CF	0.5
dbpedia:Pedro_Rios_Maestre	dbpedia:Getafe_CF	0.5
dbpedia:Lionel_Messi	dbpedia:FC_Barcelona	1.0

Table 5.4 Entity disambiguation evaluation results in the football scenario

System/approach	Precision (%)	Recall (%)	F_1 measure (%)
Proposed approach	84	81	82
AIDA	62	56	59
DBPedia Spotlight	85	26	40

Scenario 2: Resolving Locations in Military Conflict Texts

In this scenario, our task was to disambiguate location references within a set of textual descriptions of military conflicts like the following: *"The Siege of Augusta was a significant battle of the American Revolution. Fought for control of Fort Cornwallis, a British fort near Augusta, the battle was a major victory for the Patriot forces of Lighthorse Harry Lee and a stunning reverse to the British and Loyalist forces in the South."* For that we again used DBpedia as well as an evidence model based on the disambiguation evidence mapping function of Table 5.1. Using this model, we applied, as in the football scenario, our disambiguation process in a set of 50 military conflict texts, targeting the locations mentioned in them. Table 5.5 shows the achieved results.

Table 5.5 Entity disambiguation evaluation results in the military conflict scenario

System/approach	Precision (%)	Recall (%)	F_1 measure (%)
Proposed approach	88	83	85
DBPedia Spotlight	71	69	70
AIDA	44	40	42

Table 5.6 Geographical scope resolution evaluation results

System/approach	Precision (%)	Recall (%)
Proposed approach	78	76
Yahoo! Placemaker	30	30

Scenario 3: Resolving the Geographical Scope of Military Conflict Texts

In this scenario, our task was to determine the geographical scope of textual descriptions of military conflicts of the previous scenario. Using an evidence model based on Table 5.1 again we applied the thematic resolution process to a set of 100 military conflict texts and measured its effectiveness using precision and recall metrics as described above. As a baseline, we compared our results to those derived from Yahoo! Placemaker[1] geo-parsing Web services. As one can see from Table 5.6 our approach yields significantly better results.

Scenario 4: Resolving the Scope of Film Reviews

In this scenario, our task was to analyse texts containing film reviews and identify, through our method, the film that each review was about. For that purpose we built a thematic scope evidence model using an appropriate ontology, derived from Freebase. The ontology comprised about 148,000 films, 145,000 actors, 63,000 characters and the relations between them (film with directors, films with actors and films with characters). Using this model, we applied the thematic resolution process in a set of 1000 film reviews, randomly selected from a set of 25,000 IMDB reviews.[2] As a baseline, we considered a frequency-based approach where we assumed that the most frequent film within the text is the one the text talks about. Results are shown in Table 5.7.

[1] http://developer.yahoo.com/geo/placemaker/.

[2] http://www.cs.cornell.edu/people/pabo/movie-review-data.

Table 5.7 Film scope resolution evaluation results

System/approach	Precision (%)	Recall (%)
Proposed approach	89	85
Frequency-based baseline	45	45

5.2 Open-World Schema Learning for Knowledge Graphs*

In this section, we describe a way to deal with the schema learning problem from *incomplete* Web data. As we know, ontology TBoxes, or conceptual schemas, are the backbone of Knowledge Graphs, but they are always difficult to obtain, especially when the data is incomplete. In our approach, the TBox learning task in a Description Logic (DL) is transformed into a Bayesian inference task in an extension of the Bayesian Network, which is based on the original DL ontology. Bayesian Description Logic Network (abbreviated as BelNet), integrating the probabilistic inference capability of Bayesian Networks with the logical formalism of DL ontologies, supports promising inference, even when the dataset is incomplete. In this section, we first introduce the motivation for this work, explain the details of BelNet$^+$ and, finally, introduce a TBox learning approach with BelNet$^+$ based on Open World Assumption (OWA). In order to showcase the performance of this approach, a novel evaluation framework with incomplete data will be adopted to conform to the open, dynamic and non-consistent Web environment. Finally, the result from empirical studies on comparisons with the state-of-the-art TBox learners is provided, verifying the effectiveness of our approach.

5.2.1 Motivation

Ontologies are basic building blocks of the Semantic Web [121, 186]. The number of Semantic Web datasets has approximately doubled since 2011, and it has grown by 270 % if social networking is taken into account. However, the knowledge acquisition bottleneck has resulted in inexpressive schemata (also known as TBoxes, while the data part of ontologies are called ABoxes) on the Semantic Web [52, 125].

One way of enriching TBoxes is to (semi-)automatically learn TBoxes, both from unstructured documents [52] and semi-structured documents [164]. Given the fast development of semantic data, one way of exploiting it is to learn TBoxes from semantic data (ABoxes) [262]. However, in an environment like the Semantic Web, data generally suffers from incompleteness [263], which consequently hinders the learners from getting correct results. In this section, we focus on learning the TBox from incomplete ABox data.

This problem has attracted the attention of both Machine-Learning and data mining domains. For example, a number of studies have applied Inductive Logic

Programming (ILP) to learn Description Logic (DL) knowledge bases. Lehmann et al. [140] extensively studied the properties of $\mathcal{ALC}$ (a basic DL) and $\mathcal{EL}$ (a lightweight language) refinement operators, which were used in the ILP algorithm. Since the refinement operators are designed to traverse the possible candidates, the approach is effective over complete data. However, the candidate scores are based on both positive and negative examples by making a Closed World Assumption (CWA), i.e. assuming true only the specified and derivable statements; under the incomplete Semantic Web data this leads to lots of noisy negative examples.

Consequently, a candidate concept that best describes the other but is overspecialized will be selected at last. For example, one might learn an axiom $Grandson \sqsubseteq Male \sqcap \neg Person$ (*Grandson* is a *Male* who is not a *Person*) from a dataset without statements like "individual grandsons are person." In the data mining domain, Völker and Niepert [247] used association rule mining to learn TBox from the Semantic Web knowledge base such as DBpedia. The measures used to select candidate TBox axioms are support and confidence, where negative examples are out of consideration but are undoubtedly useful in specialising the axioms and decreasing the redundancies in the results. Furthermore, TBox axioms are learned for respective and independent targets, which lead to either over- or underspecialised result sets. Finally, the metrics precision, recall and F1-measure commonly used by current approaches are sensitive to minor changes in the gold standard ontologies. For example, consider a set containing "$Father \sqsubseteq Male$." Replacing "$Father \sqsubseteq Male$" and "$Father \sqsubseteq \exists hasChild.\top$" with "$Father \sqsubseteq \text{Male} \sqcap \exists \text{hasChild}.\top$" will decrease recall from 1/2 to 0. To summarise, the problem of learning TBox from incomplete Semantic Web data remains challenging because:

- Little attention is paid to approaches dealing with the incompleteness in the data.
- An evaluation framework to compare existing approaches is lacking.

In order to address the above challenges we make the following four contributions in this section.

- We generate the negative examples according to the CWA in a manner similar to that mentioned in [140]. However, to solve the noise issue brought by CWA and incompleteness, we adopt an approach that instead of merely considering the instances of concept pairs, uses also inference in a Bayesian network that leverages the structure in the data.
- In order to foster promising inference on subsumption and disjointness axioms, we extend BelNet [263] to BelNet$^+$. BelNet combines Bayesian networks with DLs by representing DL concepts as nodes and subsumptions with links. In BelNet$^+$, we extend the semantics of links in BelNetly using additional links for disjointness. Compared to BelNet, BelNet$^+$ is more effective in detecting disjoint concepts and answering queries.
- We consider the TBox learning as instance classification. In order to conform to the Open World Assumption (OWA) generally made in the Semantic Web, we extend the traditional confusion matrix by considering unknown results (neither true nor false), and propose the metrics using the new confusion matrix correspondingly.

Our extension of traditional evaluation metrics reflects more objectively on the performance of the learners.
- In order to evaluate the state-of-the-art TBox learners, we set up gold standard ontologies correspondingly. Meanwhile, in our evaluation framework, the quality of the gold standard ontologies is more easily guaranteed.

In the rest, we first introduce the BelNet$^+$ model and the TBox learning with it. Second, we describe an evaluation framework for TBox systems and show empirical performance evaluations. We also briefly review the related works and discuss the future research.

5.2.2 *BelNet$^+$*

The Syntax

Definition 1 A *Bayesian subsumption axiom* is in the form of $D \mid C_1, \ldots, C_n$, where $C_i \sqsubseteq D$, $C_i \not\equiv \bot$, $D \not\equiv \bot$, $i \in \{1, \ldots, n\}$ and $\nexists j, k \in \{1, \ldots, n\}$ such that $C_j \sqsubseteq C_k$. If $D' \equiv D$, then label D with an *alias* D'.

Definition 2 A *Bayesian disjoint axiom* is in the form of $D \mid \overline{C}$, where $C \sqcap D \sqsubseteq \bot$ and $C \not\equiv \bot, D \not\equiv \bot$.

Definition 3 A BelNet$^+$ contains a set of Bayesian subsumption axioms $(D \mid C_1, \ldots, C_n)$ and a set of Bayesian disjoint axioms $(D \mid \overline{C})$, together with an ontology ABox. A BelNet$^+$ defines a Bayesian network $\mathcal{B}$ as follows:

- $\mathcal{B}$ contains one binary node associated with a conditional probability table (CPT) calculated from the ABox for each C_i and D appearing in either the Bayesian subsumption axioms or the Bayesian disjoint axioms.
- $\mathcal{B}$ links from node C_i to node D for each Bayesian subsumption axiom $D \mid C_1, \ldots, C_n$, $i \in \{1, \ldots, n\}$.
- There is a link between node C and node D in each Bayesian disjoint axiom.

Links in a BelNet$^+$ can be *conditional*, which means that the assignments of one node fully determine that of the other one. Figure 5.1 shows an example of conditional links: The links from *Male* and *Female* to *Female*$\sqcup$*Male* are conditional. With the assignments for variables *Female* and *Male*, we know for sure the assignment of *Female* $\sqcup$ *Male* by the semantics of DLs.

For convenience, in this section we use the same symbol for both the concept in DL ontology and the corresponding node in the Bayesian network.

Example 1 Given an ontology containing TBox $\{Male \sqsubseteq Person, Female \sqsubseteq Person, Male \sqcap Female \sqsubseteq \bot\}$, the corresponding Bel Net$^+$ contains the following Bayesian subsumption axioms and Bayesian disjoint axioms:

$$Person \mid Male, Female$$
$$Male \mid \overline{Female}$$

this BelNet^{+} specifies a Bayesian network structure as shown in Fig. 5.2, where the CPTs are learned by parameter learning (Sect. 5.2.2).

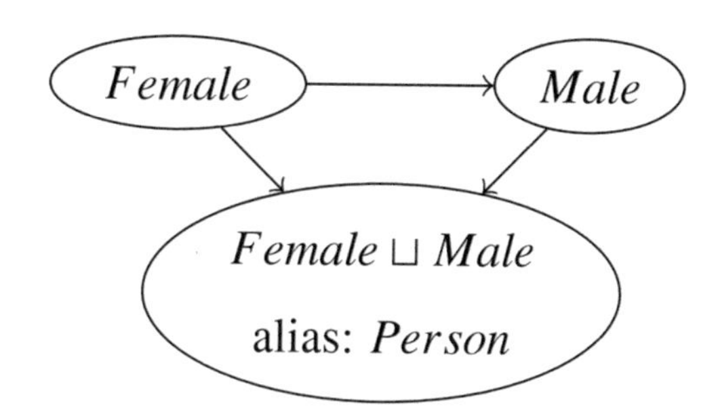

Fig. 5.1 An example of conditional links

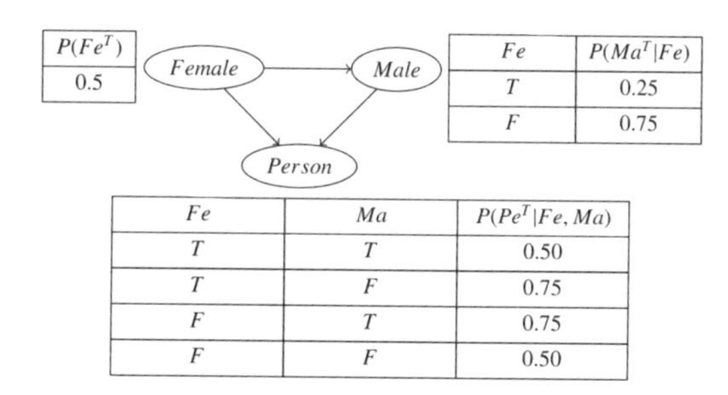

$P(Fe^T)$
0.5

Fe	$P(Ma^T \mid Fe)$
T	0.25
F	0.75

Fe	Ma	$P(Pe^T \mid Fe, Ma)$
T	T	0.50
T	F	0.75
F	T	0.75
F	F	0.50

Fig. 5.2 A motivated BelNet^{+} example. *Ma*, *Fe* and *Pe* are short for *Male*, *Female* and *Person*. T and F are short for values `TRUE` and `FALSE`

In fact, we can prove that a BelNet^{+} is guaranteed to define a Bayesian network as a directed acyclic graph (DAG).

Semantics

The semantics of a BelNet^{+} is based on joint probability distributions over the Bayesian network generated as follows:

$$P(\mathcal{B}) = P(C_1, \ldots, C_n) = \prod_{i=1}^{n} P(C_i \mid Pa(C_i)) \tag{5.7}$$

where C_is are nodes in $\mathcal{B}$, and $Pa(C_i)$ is the parent set of C_i.

Example 2 By calculating from the BelNet^{+} shown in Fig. 5.2, the joint probability of the existence of an instance who is a Female, a Male, and a Person is $0.5 \times 0.25 \times 0.50 = 0.0625$. The probability of an instance who is a Female and also a Person is $0.5 \times 0.25 \times 0.75 + 0.5 \times 0.75 \times 0.75 = 0.375$. In this example,

($P(Person^F \mid Female^T, Male^F) = 0.25$), the probabilities calculated still suggest that the first instance is less probable.

A BelNet$^+$ can be viewed as a template for generating ABoxes. Given different sets of conditional probability tables (or CPTs), or different set of Bayesian axioms, it can generate different ABoxes.

Parameter Estimation

The parameters in a BelNet$^+$ refer to the CPT in the Bayesian network defined by it. In this part, we will discuss how the parameters can be learned from Semantic Web data.

It is natural to use a finite ontology domain $\Delta^{\mathcal{I}}$ to restrict all individuals in the ABox of a BelNet$^+$. We assume that $\Delta^{\mathcal{I}}$ contains all individual names in the BelNet$^+$, and an individual name o is always interpreted to itself, i.e. $o^{\mathcal{I}} = o$.

We call all interpretations related to individual o a *possible observation* $\mathbf{o}$. For example, $C^{\mathcal{I}} = \{a, b\}$, then there are two possible observations, where $C^{\mathbf{o}_1} = \{a\}$, $C^{\mathbf{o}_2} = \{b\}$. A possible observation is an interpretation which assigns at most one element to one concept. Actually, under a possible observation, C_i has two values: T and F. For a specific observation $\mathbf{o}$, $\mathbf{o}$ supports C_i^{T} if $o \in C_i^{\mathcal{I}}$, and $\mathbf{o}$ supports C_i^{F} if $o \notin C_i^{\mathcal{I}}$. These cases can be abbreviated as $C_i^{\mathbf{o}}$. $\mathcal{B}^{\mathbf{o}}$ is short for $\{C_1^{\mathbf{o}}, \ldots, C_n^{\mathbf{o}}\}$, where $C_i, i \in \{1, \ldots, n\}$ is a node in $\mathcal{B}$.

For a *marginal node* C, which has no parents in $\mathcal{B}$, the *marginal probability* is a table of $P(C^{\sharp})$, where $\sharp \in \{\mathtt{TRUE}, \mathtt{FALSE}\}$. Furthermore, $P(C^{\mathtt{TRUE}})$ is the probability that a possible observation supports C, i.e. $P(o \in C^{\mathcal{I}})$. Similarly $P(C^{\mathtt{FALSE}})$ is the probability that a possible observation does not support C, i.e. $P(o \notin C^{\mathcal{I}})$. Actually the parameters depend on the number of individuals satisfying concept C in the ontology, as we will see below. For convenience, in the following TRUE(/FALSE) is shortened to be T(/F).

The CPTs will be learned from the ontology ABox. We assume that all possible observations are independent. By Eq. (5.7) the likelihood of all possible observations $\{\mathbf{o}\}$ is

$$L(\theta : \{\mathbf{o}\}) = \prod_{\mathbf{o}} \prod_{i=1}^{n} \theta_{C_i^{\mathbf{o}} \mid Pa(C_i)^{\mathbf{o}}} = \prod_{i=1}^{n} \theta_{C_i^{\mathbf{o}} \mid Pa(C_i)^{\mathbf{o}}}^{N[C_i^{\mathbf{o}} \mid Pa(C_i)^{\mathbf{o}}]} \tag{5.8}$$

where θ denotes the set of CPT values and $N[C_i^{\mathbf{o}} \mid Pa(C_i)^{\mathbf{o}}]$ is the number of possible observations satisfying $C_i^{\mathbf{o}} \mid Pa(C_i)^{\mathbf{o}}$. Maximising this likelihood by setting the derivative of the log-likelihood of Eq. (5.8) with respect to its CPTs to 0 results in

$$\theta_{C_i^{\mathbf{o}} \mid Pa(C_i)^{\mathbf{o}}} = \frac{N[C_i^{\mathbf{o}} \mid Pa(C_i)^{\mathbf{o}}]}{N[Pa(C_i)^{\mathbf{o}}]} \tag{5.9}$$

In order to avoid the cases where $N[Pa(C_i)^{\mathbf{o}}] = 0$, we add one "imaginary" possible observation to it.

Example 3 Given the BelNet$^+$ in Example 1. In addition, we also have an ABox:

$$Person(a), Person(b), Male(a), Female(b)$$

Then the estimation of $\theta_{Person^T \mid Female^T, Male^F}$ is $\frac{1+0.5}{1+1} = 0.75$.

The learned CPTs are shown in Fig. 5.2.

Inference

BelNet$^+$ can answer an arbitrary probability query: "Given a BelNet$^+$, what is the probability of a Bayesian subsumption / disjoint axiom?." More formally, the conditional probability query is given by

$$P(D \mid C_1, \ldots, C_n) = \frac{P(\sqcup_{i=1}^{n} C_i^T, D^T)}{P(\sqcup_{i=1}^{n} C_i^T)} \tag{5.10}$$

and

$$P(D \mid \overline{C}) = 1 - P(D^T, C^T) \tag{5.11}$$

Equations (5.10) and (5.11) can be calculated by joint probabilities over the Bayesian networks. Joint probability queries can be answered in Bayesian networks. In our implementation, we use the junction tree algorithm [89] to do the task.

Structure Learning

Structure learning is a specific type of knowledge discovery that learns a dependency structure, being able to give promising answers to queries "what is the probability of a Bayesian subsumption/disjoint axiom?." So the task of *structure learning* in BelNet$^+$ is to find a BelNet$^+$ $\mathcal{B}$ that makes the data the most probable. This is similar to the task of structure learning in Bayesian networks except that the structure we learn needs to be a BelNet$^+$. In other words, the links in the structure need to be corresponded to subsumption or disjoint relationships. If we denote the candidate structures in a domain as $\mathcal{B}^+$, and that of the same domain in Bayesian networks as $\mathcal{B}$, we have $\mathcal{B}^+ \subseteq \mathcal{B}$. Thus, we can share the structure scores from that in Bayesian network structure learning.

Structure Score. Choices for score functions used in Bayesian network structure learning include maximum likelihood, Bayesian score that is based on a Bayesian perspective encoding uncertainties both over structure and over parameters, and extensions of Bayesian score. Likelihood measurement suffers from overfitting, and prefers more complex networks to simpler ones, which is not always the preference in practice. For handling overfitting problems and more efficient numerical computation of the Bayesian score [136], we will adopt the decomposable Bayesian score with Dirichlet priors as our score function.

Structure Search. We knew from literature that "Given a dataset $\mathcal{D}$ and a decomposable score function, finding $\mathcal{G}^* = \arg\max_{\mathcal{G}\in\mathcal{G}_d} score(\mathcal{G} : \mathcal{D})$ is $\mathcal{NP}$-hard." [136]. The BelNet$^+$ structure would additionally have the property that the links correspond to subsumption or disjoint relationships. Thus, instead of aiming for an algorithm that will always find the highest-scoring network, we resort to heuristic algorithms that attempt to find the best network but are not guaranteed to do so. The algorithm adopted here is a modified version of the structure learning algorithm in Bayesian networks [173]. The Bayesian network structure learning algorithm can only recover the structure that is equivalent in terms of representing the independencies among the nodes to the real underlying structure [136]. In this section, the preference is a single structure that is concise and can directly be used to extract axioms. To achieve this goal, we incorporate this preference in our algorithm.

Roughly speaking, the structure learning algorithm starts from an initial structure (with nodes, and the conditional links between nodes), and iteratively tries to find the best operation (in terms of adding/deleting/reversing) that can be carried out from the current structure, unlike in [263]. This process iterates until no better structure can be found, or the step reaches the maximum threshold. Two thresholds are involved in this procedure. One controls the maximum number of parent nodes a node can have, the other controls the maximum number of iterations for this procedure to exit.

Selection criteria. After an operation is selected by the score function, in order to meet the demand of BelNet$^+$, to be specific, preference is given to structures whose links signify the special dependencies "subsumption" and "disjointness," different from [263]. The operations not satisfying the requirements are filtered out by the selection criteria.

We denote the candidate operation as *op*, where op_{head} is the node to which the link points, and op_{tail} represents the node from which the link starts. Further, we denote the count of instances belonging to concepts op_{tail} and op_{head} as $M[op_{head}, op_{tail}]$, the count of instances belonging to concept op_{head} as $M[op_{head}]$, similar for $M[op_{tail}]$. Then, operation *op* will be selected iff either $M[op_{head}, op_{tail}] = M[op_{tail}]$ *and* $M[op_{tail}] > threshold_{parent}$ *or* $M[op_{head}, op_{tail}] = 0$ *and* $M[op_{head}] \neq 0$ *and* $M[op_{tail}] \neq 0$. In the experiment, the $threshold_{parent}$ is set as 0.

It happens that some concepts in the ontology contain a large number of missing values. Those corresponding nodes are out of consideration in the post-processing step. Rest of the nodes are called *informative nodes*.

In our algorithm, besides Bayesian disjoint axioms, which are considered in [263] on a smaller scale, the candidate Bayesian subsumption axioms are also generated by inference over B, and the results of the inference can be considered as weights of the candidates. In practice, in order to select the axioms from the weighted results, we use thresholds. Since the Bayesian network constructed can behave differently for Bayesian subsumption axioms and Bayesian disjoint axioms, we use $threshold_{subsumption}$ and $threshold_{disjoint}$, respectively, for the selection.

Although the results could be quite simple, such as relations between the pair of concepts generated from pre-processing, these are however the basis for more complex axioms, as shown below.

- If there is more than one Bayesian subsumption axiom $\{D_1 \mid C, D_2 \mid C, \ldots, D_n \mid C\}$, generate $C \sqsubseteq \sqcap_{i \in \{1,\ldots,n\}} D_i$.
- If there is more than one Bayesian subsumption axiom $\{D \mid C_1, D \mid C_2, \ldots, D \mid C_n\}$, generate $\sqcup_{i \in \{1,\ldots,n\}} C_i \sqsubseteq D$.
- Bayesian disjoint axioms correspond to disjoint axioms in ontologies.

5.2.3 *TBox Learning as Inference*

After describing the details of BelNet$^+$, we will introduce how the TBox can be learned with BelNet$^+$. The learning approach includes three main steps:

1. *Pre-processing*. In pre-processing, given an ontology $\mathcal{O}$, for each $C \in N_C^+$ and $r \in N_R$, pre-processing creates nodes corresponding to C and $\exists r.\top$. Conditional links are added among the nodes. Furthermore, ABox materialisation will be carried out on each node generated in this step. We denote the ABox materialised ontology as $\mathcal{O}^+$. The result of this step is denoted by $\mathcal{B}^0$.
2. *Learning Bayesian network*. Structure learning (cf. Sect. 5.2.2) will be carried out on $\mathcal{B}^0$ over $\mathcal{O}^+$. After that, the parameter learning will fill the CPTs attached with the structure learned. We denote the result of this step as $\mathcal{B}$.
3. *Post-processing*. Having a Bayesian network learned, TBox axioms are extracted through inference over $\mathcal{B}$. See below for details.

5.2.4 *A Novel Evaluation Framework*

The set of axioms learned by TBox learning systems can be viewed as an application of Information Retrieval on a knowledge base. From this perspective, the performance of a TBox learning system can either be judged by human experts or be evaluated by traditional IR measures. Using traditional IR measures, an axiom learned is *correct* if it can be entailed by the gold standard ontology. However, both methodologies suffer from disadvantages: human experts are subjective to some extent, and there are various representations for a domain, consequently, the evaluations by IR measures are sensitive to gold standard ontologies.

From another perspective, the TBox in an ontology assists in classifying instances with DL reasoners. Although it is impossible to explicitly make all true statements of the interested domain, it is still workable to get as many facts as possible through reasoning. In this way, the TBox can be viewed as a set of classification "rules" to classify the instances. Based on this observation, we extend metrics used in the classification.

Below, we first introduce the notations we will use in the evaluation framework.

Notations. We denote the original ontology (the input of ontology learners) as $\mathcal{O}$, and the output as $\mathcal{O}'$. Furthermore, the gold standard ontology is denoted by $\mathcal{O}^S$.

Definition 4 (*Gold standard ontology*) An ontology $\mathcal{O}^S$ is called a gold standard ontology for $\mathcal{O}$, if $\mathcal{O}^S$ satisfies the following:

- $\mathcal{O}^S$ is both consistent and coherent.
- $\mathcal{O}^S$ entails all correct (with respect to the knowledge of domain experts) ABox statements with the vocabulary of its ontology counterpart $\mathcal{O}$.

Property 1 *The gold standard ontology $\mathcal{O}^S$ of an ontology $\mathcal{O}$ can be non-unique.*

Property 1 is straightforward. A gold standard ontology for $\mathcal{O}$ can be the one with ABox knowledge not explicitly stated but inferred. In the extreme case, another gold standard ontology for $\mathcal{O}$ may explicitly state all ABox statements.

If we view the TBox as a set of classification rules, the result of classifying an instance a towards a concept A with respect to an ontology $\mathcal{O}$ is

$$f(a, A, \mathcal{O}) = \begin{cases} \text{positive} & \mathcal{O} \models A(a) \\ \text{negative} & \mathcal{O} \models \neg A(a) \\ \text{unknown} & \text{otherwise} \end{cases}$$

In order to incorporate the *unknown* values in the classification results, we extend the traditional confusion matrix used in the evaluation of binary classification [108] by considering “unknown” as a specific classification result (cf. Table 5.8).

With this extension in hand, several classical metrics used by classification problems are (extended) as follows:

$$Accuracy(U) = \frac{TP + TN + w \cdot TU}{P_C + N_C + w \cdot U_C}$$

$$ErrorRate(U) = 1 - Accuracy(U)$$

$$Precision(U) = \frac{TP}{TP + FP(N) + w \cdot FP(U)}$$

Table 5.8 The extended confusion matrix

	P_C	N_C	U_C
P	TP	FP(N)	FP(U)
N	FN(P)	TN	FN(U)
U	FU(P)	FU(N)	TU

T and *F* are short for *True* and *False*. *P*, *N* and *U* are short for *Positive*, *Negative* and *Unknown*, respectively. *FP*(*N*) is short for *False Positives* from *Negatives* (set of positive results which should be labelled as negatives)

$$Recall(U) = \frac{TP}{TP + FN(P) + w \cdot FU(P)}$$

$$F\text{-}Measure(U) = \frac{(1+\beta)^2 \cdot Recall(U) \cdot Precision(U)}{\beta^2 \cdot Recall(U) + Precision(U)}$$

$$TP\_rate = \frac{TP}{P_C}$$

$$FP(N)\_rate(U) = \frac{FP(N)}{N_C}$$

$$FP\_rate(U) = \frac{FP(U) + FP(N)}{N_C}$$

Traditional *Accuracy*, *ErrorRate*, *Precision*, *Recall* and *F-Measure* are calculated from the extended metrics when w is 0. The traditional ROC graph is formed by plotting TP_rate over $FP(N)\_rate(U)$.

We demonstrate the necessity of this confusion matrix extension by Example 4:

Example 4 Table 5.9 shows three ontologies. The first one is the gold standard ontology, where $\mathcal{O}_1$ and $\mathcal{O}_2$ are two ontologies which have to be evaluated.

If we calculate the accuracy in classifying the concept Female, then using the traditional confusion matrix,

$$Accuracy(\mathcal{O}_1, \text{Female}, \mathcal{O}^S)$$
$$= Accuracy(\mathcal{O}_2, \text{Female}, \mathcal{O}^S) = \frac{2+1}{4}$$

but apparently $\mathcal{O}_2$ contains one incorrect subsumption axiom. In the new framework, if $w = 1$, then

$$Accuracy(\mathcal{O}_1, \text{Female}, \mathcal{O}^S) = \frac{2+1+1}{4}$$

$$Accuracy(\mathcal{O}_2, \text{Female}, \mathcal{O}^S) = \frac{2+1}{4}$$

Table 5.9 An example of gold standard ontology and test ontologies

	$\mathcal{O}^S$	$\mathcal{O}_1$	$\mathcal{O}_2$
TBox	Female⊓Male⊑ ⊥	Mother⊑ Female Daughter⊑ Female Female⊓Male⊑ ⊥	Mother⊑Female Daughter⊑ Female Female⊓Male⊑ ⊥ Child⊑ Daughter
ABox	Female(a), Female(b) Male(c), Child(d)	Mother(a), Daughter(b) Male(c), Child(d)	Mother(a), Daughter(b) Male(c), Child(d)

All ontologies have the same set of concept names: Female, Male, Mother, Daughter and Child

The measures mentioned above are used in binary classification evaluations. When evaluating a multi-class problem, we simply use an average (weighted) of the above measures. Suppose the importance of the concepts is ranked with weights $w_1, \ldots, w_n$, then the average (weighted) value of a specific measure is

$$\frac{\sum_{i=1}^{n} w_i \cdot Measure(A_i)}{\sum_{i=1}^{n} w_i}$$

where *Measure* can be replaced by any of the (extended) metrics listed above. A_i denotes the ith concept.

Property 2 *The (extended) metrics listed above are the same in all gold standard ontologies* $\mathcal{O}^S$.

Property 2 holds because the (extended) metrics are calculated by the extended confusion matrix (Table 5.8), and according to the definition of gold standard ontologies, the confusion matrix is the same in all gold standard ontologies.

Property 2 indicates the stability of a gold standard ontology. In other words, the variations in gold standard ontologies have no influence on the evaluation results.

5.2.5 Experiments

We have implemented a prototype of BelNet$^+$ with the TBox learning algorithm in Java and Scala. We designed and carried out the experiments to highlight the effect of incompleteness on learning methods. In this section, we evaluate the performance of the proposed learning method focusing on answering the following three questions: (1) How promising is the inference in BelNet$^+$? (2) How are the performances of the four approaches, namely DLLearner, GoldMiner, BelNet and BelNet$^+$, under the existence of incompleteness? (3) Will the amount of incompleteness be decreased with TBox learning?

Experimental Setup: Dataset

The datasets used in the experiments include the following: Family,[3] Semantic Bible (NTN),[4] LUBM[5] and Wine.[6] We manually constructed gold standard ontologies for the datasets.[7]

[3]https://github.com/fresheye/belnet/blob/master/ontology/family-benchmark_rich_background.owl.

[4]http://www.semanticbible.com.

[5]http://swat.cse.lehigh.edu/projects/lubm/.

[6]http://kaon2.semanticweb.org.

[7]https://github.com/fresheye/belnet/blob/master/ontology/.

In order to quantify the degree of incompleteness of an ontology $\mathcal{O}$, we denote incompleteness by the percentage of unknown answers to all possible queries in the form "Is individual a an instance of concept A?"

To be specific, the *incompleteness* of an ontology $\mathcal{O}$ is quantified by

$$\frac{|\{f(a, A, \mathcal{O} \mid f(a, A, \mathcal{O}) \text{ is unknown}\}|}{|\{f(a, A, \mathcal{O}\})|}$$

where $a \in N_I$ and $A \in N_C$.

The relevant statistics of the datasets and the corresponding gold standard ontologies are shown in Table 5.11, in which we calculate the number of named concepts, object properties, number of subClassOf, equivalentClass, disjointWith axioms, number of individuals, the DL expressibility and the incompleteness of the corresponding ontologies. As shown in the table, Semantic Bible, LUBM and Wine contain more incompleteness than Family (Table 5.10).

It is worth noticing that the DL expressibility of the ontologies chosen is not restricted to certain DL languages. In our approach, all concept expressions in the original ontology are treated as concepts.

To demonstrate TBox learner's capability in handling incompleteness, we create subontologies of the original ontologies with different levels of incompleteness. We partition the ABox into 10 parts. Then, we randomly select one of them and add it to the TBox as the first subontology. By randomly selecting and adding one part to the existing largest subontology each time, we finally get 10 subontologies. This procedure is carried out 10 times, each with a different initial start subontology. In order to clearly demonstrate the performance, the result ontologies only contain the axioms learned.

Default Values and Thresholds

Goldminer consists of four tunable parameters, namely support and confidence in learning subsumptions and disjointness separately. We tried parameters in the scope of $[0 - 1]$ for Goldminer, and finally we chose the support threshold to be 0, and the confidence threshold to be 0.9 for learning subsumptions, and 0.1 (support), 0.8 (confidence) for learning disjointness, which is also the setting recommended in [79], in order to get a higher F-measure.

In BelNet$^+$, we tried different combinations of the maximum number of parents and maximum number of iterations. We set 5 as the maximum number of parents and 100 as the maximum number of iterations, because the results are almost stable with these settings. In addition, we only learn axioms among the concepts containing at least 10 % of individuals. The corresponding concepts for informative nodes contain at least one individual.

Table 5.10 The thresholds and AUC of each partition per dataset (t_d: $threshold_{disjoint}$, AUC_d: $AUC_{disjoint}$, t_s: $threshold_{subsumption}$, AUC_s: $AUC_{subsumption}$)

%	Family				NTN				LUBM				Wine			
	t_d	AUC_d	t_s	AUC_s	t_d	AUC_d	t_s	AUC_s	t_d	AUC_d	t_s	AUC_s	t_d	AUC_d	t_s	AUC_s
10	0.9900	0.8044	0.4082	0.9126	0.9982	0.9128	0.8670	0.9931	0.9833	0.9210	0.8657	1.0000	0.9811	0.9210	0.9349	1.0000
20	0.9900	0.9348	0.3729	0.9902	0.9976	0.8790	0.6053	0.9886	0.9861	0.9374	0.9176	1.0000	0.9902	0.9374	0.9562	1.0000
30	0.9935	0.9756	0.8869	0.9995	0.9982	0.8834	0.3630	0.9872	0.9903	0.9282	0.9198	1.0000	0.9919	0.9282	0.9643	1.0000
40	0.9945	0.9899	0.9263	0.9995	0.9980	0.9020	0.6444	1.0000	0.9805	0.9555	0.9353	1.0000	0.9936	0.9555	0.9848	1.0000
50	0.9963	0.9981	0.9486	1.0000	0.9982	0.9249	0.7477	0.9999	0.9866	0.9542	0.9277	1.0000	0.9902	0.9542	0.9873	1.0000
60	0.9966	0.9992	0.9581	1.0000	0.9991	0.9589	0.8311	0.9996	0.9728	0.9509	0.9435	1.0000	0.9880	0.9509	0.9796	1.0000
70	0.9961	0.9997	0.9595	1.0000	0.9997	0.9538	0.8154	0.9987	0.9672	0.9561	0.9582	1.0000	0.9879	0.9561	0.9814	1.0000
80	0.9961	0.9998	0.9589	1.0000	0.9994	0.9381	0.3417	0.9945	0.9821	0.9650	0.9671	1.0000	0.9904	0.9650	0.9875	1.0000
90	0.9956	1.0000	0.9549	1.0000	0.9987	0.9650	0.6390	0.9998	0.9551	0.9591	0.9738	1.0000	0.9935	0.9591	0.9876	1.0000
100	0.9913	1.0000	0.9569	1.0000	0.9972	0.9748	0.9550	1.0000	0.9985	0.9459	0.9987	1.0000	0.8945	0.9459	0.8765	1.0000

If the thresholds under constraint $FPR < 0.1$ and $TPR > 0.7$ are not available, we relax the constraint of FPR by 0.1

Table 5.11 Statistics of the datasets for evaluation

Ontology	# concepts	# object properties	# $\sqsubseteq$ / $\equiv$ / $\bot$	# individuals	DL expressibility	Incompleteness
Family	19	4	27 / 0 / 0	202	$\mathcal{AL}$	0.609
Family′	19	4	27 / 17 / 14	202	$\mathcal{ALC}$	0.267
Semantic Bible	49	29	51 / 0 / 5	724	$\mathcal{SHOIN}(\mathcal{D})$	0.887
Semantic Bible′	49	29	52 / 6 / 34	724	$\mathcal{SHOIN}(\mathcal{D})$	0.048
LUBM	43	25	36 / 6 / 0	1555	$\mathcal{ALEHI}(\mathcal{D})$	0.946
LUBM′	43	25	36 / 6 / 52	1555	$\mathcal{SHI}(\mathcal{D})$	0.097
Wine	142	13	126 / 61 / 1	162	$\mathcal{SHOIN}$	0.957
Wine′	142	13	186 / 61 / 21	162	$\mathcal{SHOIN}$	0.197

The dataset name end with ′s is the gold standard dataset

To set threshold$_{disjoint}$ and threshold$_{subsumption}$ parameters, we draw the ROC (Receiver Operating Characteristic) curves for each dataset (cf. Fig. 5.3). An axiom is true if it can be entailed by the gold standard ontology, and false if not. We selected the thresholds by setting FPR (False Positive Rate) <0.1 and TPR (True Positive Rate) >0.7. The thresholds selected are shown in Table 5.10.

5.2.6 Experimental Results

Performance of Inference

The first experiment is to demonstrate the effectiveness of the inference in BelNet$^+$. For each of the four datasets, we performed the experiments by conducting two kinds of inferences, namely inference for probabilities of Bayesian subsumption axioms and Bayesian disjoint axioms, in the $\mathcal{B}$ learned.

Quality of Inference. We consider the inference results as the output of a binary classifier. By consulting the gold standard ontologies $\mathcal{O}^{\mathcal{S}}$, the correctness of the corresponding axioms can be calculated.

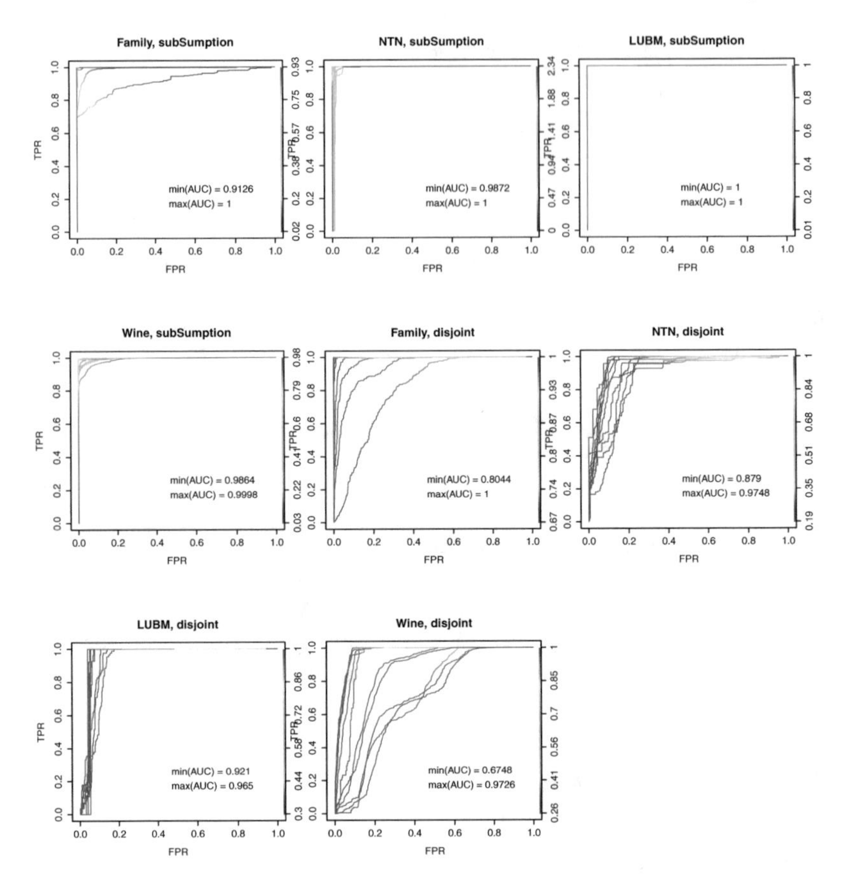

Fig. 5.3 ROC (Receiver Operating Characteristics) curves for each dataset, varying the size of the dataset (10–100 %). AUC is short for "Area Under ROC Curve"

Suppose the ontology learned is $\mathcal{O}$, *precision* and *recall* are calculated as follows:

$$Precision(\mathcal{O}^S, \mathcal{O}') = \frac{|\{\alpha \mid \alpha \in \mathcal{O}' \text{ and } \mathcal{O}^S \models \alpha\}|}{|\{\alpha \mid \alpha \in \mathcal{O}'\}|} \tag{5.12}$$

$$Recall(\mathcal{O}^S, \mathcal{O}') = \frac{|\{\alpha \mid \alpha \in \mathcal{O}^S \text{ and } \mathcal{O}' \models \alpha\}|}{|\{\alpha \mid \alpha \in \mathcal{O}^S\}|} \tag{5.13}$$

where α is a subsumption or disjointness axiom. *F-measure* is the harmonic mean of precision and recall. In Fig. 5.3, we report the inference quality by drawing the ROC curves on each partition of the four datasets. We find that:

- The inference of Bayesian subsumption axioms obtains better results than that of Bayesian disjoint axioms. As we can find from Sect. 3.7, the probability of a Bayesian subsumption axiom is a normalised measure. However, the probabilities of Bayesian disjoint axioms depend on probability queries like $P(C^T, D^T)$. On

Semantic Web, the number of individuals belonging to a pair of concepts is not large enough, which deviates the results.

- The areas under the ROC curve (AUCs), a.k.a. the probability that inference as a classifier ranks higher for correct axioms than for incorrect axioms, in the figure are quite high. Thus, the effectiveness of inference is confirmed.
- For both subsumption and disjointness axioms, the performance of inference gets better with the size of datasets growing.
- After the thresholds are selected, we compare the precision, recall and F-measure of the axioms learned by the four approaches. As shown in Table 5.12, BelNet$^+$ outperforms the other three approaches in terms of F-measure. Worth noticing is that the precision of BelNet$^+$ is always the highest in all datasets, which also confirms our expectation that BelNet$^+$ gives promising results of queries.
- From the whole dataset (cf. the rows in Table 5.12 for data partition 100 %), which is the real-world ontology, BelNet$^+$ also outperforms other learners.

Performance of Instance Classification

We now compare the performance of BelNet$^+$ with DLLearner, Goldminer and BelNet in our proposed evaluation framework.

Quality of Classification. In order to show the effect of incompleteness over learners, we partition the datasets with respect to ABox assertions. Figure 5.4 illustrates the average accuracy of classifying the instances in each dataset. Because there is no preference as to the concepts to be classified, we set equal weights to each concept. We demonstrate the average accuracies on both training sets and the whole datasets. The figures in the upper row show the average accuracy on the training sets, and figures in the row below are the average accuracy on the whole dataset. From these figures, it is not hard to find the following:

- Although the average accuracies on training sets of BelNet$^+$ are not guaranteed to be the highest, they are the highest on the whole datasets in all of the tests. It proves that BelNet$^+$ is effective in instance classification under the existence of incompleteness.
- The average accuracy of BelNet$^+$ on the whole datasets goes closer to that on the training datasets. This shows that the performance of BelNet$^+$ gets better with the size of datasets growing, which is the same conclusion as that reached in the previous sections.
- Among the four learners, BelNet behaves similarly as BelNet$^+$ in terms of trend. This is not surprising, because BelNet$^+$ is an extension of BelNet.
- The average accuracies of DLLearner and Goldminer are relatively low on Family dataset, which shows that the performance of the two learners is affected more by the incompleteness in the datasets.

Effectiveness of Incompleteness Reduction. Having verified the performance of the learning approaches, in the sequel we will evaluate the effectiveness of the learners in reducing incompleteness. Figure 5.5 represents the incompleteness of the ontologies

Table 5.12 Quality of inference for 50 and 100 % of the datasets (P: Precision, R: Recall, F: F-measure, DLer: DLLearner, Gold: Goldminer, Bel: BelNet, Bel^+: $BelNet^+$)

%		Family				NTN			
		DLLearner	Goldminer	BelNet	$BelNet^+$	DLLearner	Goldminer	BelNet	$BelNet^+$
50	Precision	0.0778	0.5455	1.0000	1.0000	0.1875	0.5193	0.6244	0.9683
	Recall	0.0074	0.9630	0.5935	0.7561	0.1512	0.7679	0.7791	0.5577
	F-measure	0.0044	0.6964	0.7448	**0.8611**	0.1674	0.6196	0.6932	**0.7077**
100	Precision	0.0556	0.5175	0.8148	0.9306	0.2500	0.6179	0.6127	1.0000
	Recall	0.2222	0.9259	0.4634	0.8293	0.1977	0.8571	0.7791	0.7791
	F-measure	0.0889	0.6640	0.5908	**0.8770**	0.2208	0.7181	0.6860	**0.8758**
%		LUBM				Wine			
		DLLearner	Goldminer	BelNet	$BelNet^+$	DLLearner	Goldminer	BelNet	$BelNet^+$
50	Precision	0.3023	0.3293	0.5347	0.8755	0.3333	0.3953	0.6700	0.4340
	Recall	0.2045	0.3611	0.4412	0.3529	0.0821	0.1498	0.1164	0.2816
	F-measure	0.2440	0.3445	0.4835	**0.5031**	0.1317	0.2172	0.1979	**0.3382**
100	Precision	0.2558	0.3474	0.5802	0.7147	0.0993	0.4765	0.6100	0.4998
	Recall	0.3409	0.3889	0.4118	0.5294	0.2657	0.3478	0.1594	0.3382
	F-measure	0.2923	0.3670	0.4817	**0.6083**	0.1446	0.4021	0.2528	**0.4034**

learned by the learners, where the incompleteness in the original ontologies is shown by a dashed line. We find that:

- Among the four learners, the performance of a DLLearner is better when the dataset is larger.
- In all datasets, BelNet and BelNet$^+$ successfully improve the completeness in the original datasets.
- Compared with all other three learners, BelNet$^+$ decreases the most incompleteness except on datasets LUBM and Wine when the partition size is relatively large. LUBM is a large dataset, making CWA still causes a DLLearner to get a large set of consistently expressive TBox axioms, which decreases the incompleteness in the dataset. On the expressive dataset Wine, the DLLearner is able to generate specific concepts when learning concept definitions. As a result, the learned axioms are effective in decreasing the incompleteness.

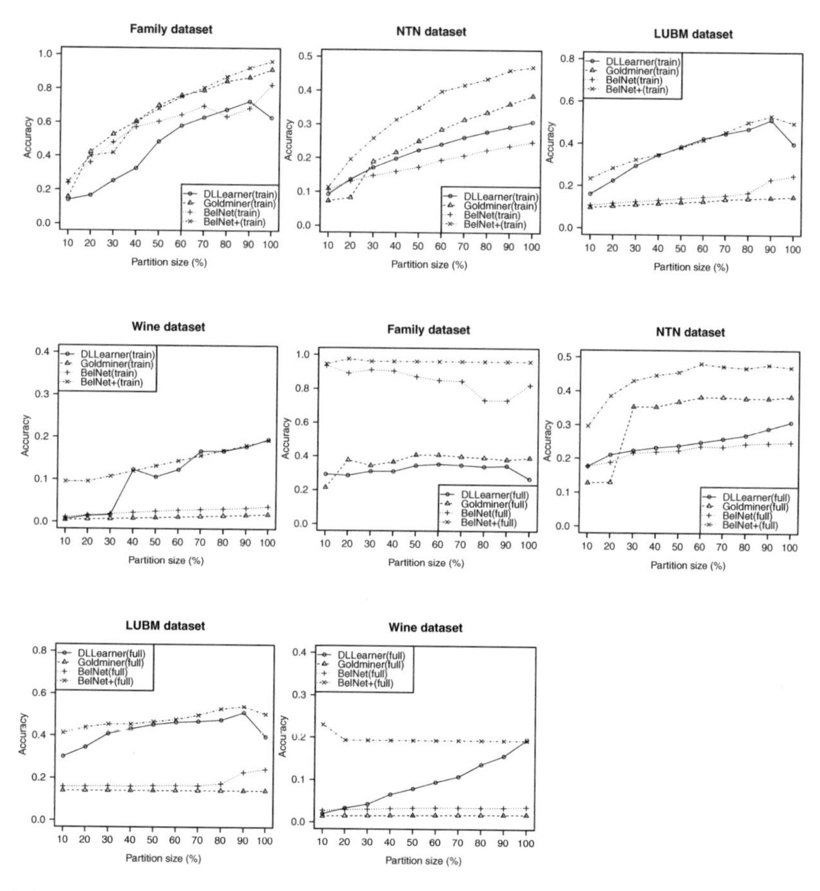

Fig. 5.4 Average accuracy of instance classification for each dataset, varying the size of the dataset. The figures in the *upper row* show the average accuracies on training sets, and figures in the *row below* show that on the whole datasets

5.2.7 *Related Work and Summary*

Since we are facing an era in which Semantic Web data grows very rapidly, learning TBox from ABox data has been attracting a lot of attention in the past five years. In this section, we notice a subset of works of ontology learning and statistical relational learning (SRL) that (1) focus on learning TBox axioms from ABox data, or (2) SRL models that handle DLs, or have applications in TBox learning.

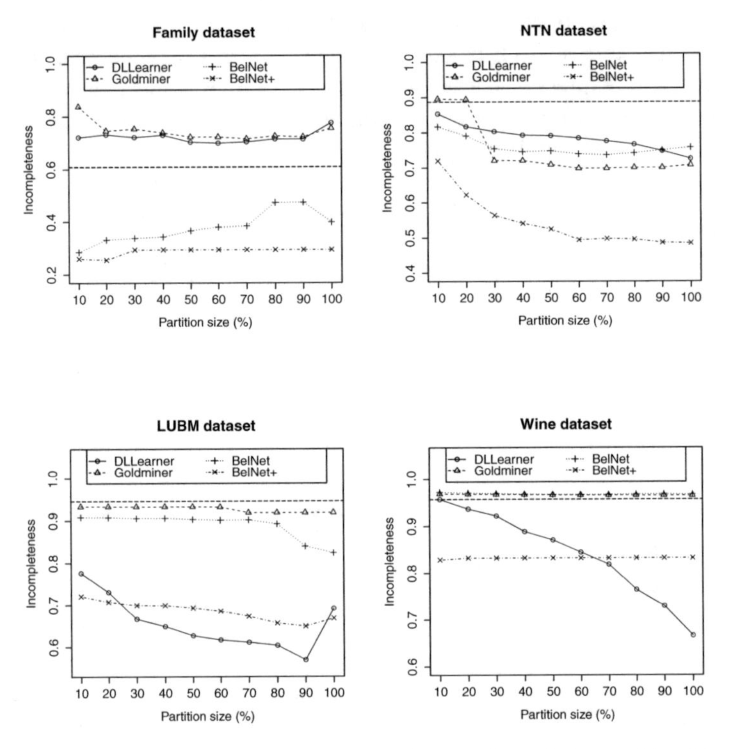

Fig. 5.5 Incompleteness in the ontologies learned by learners for each data partition. The *straight lines* in the figures indicate the incompleteness in the original datasets

Inductive Logic Programming. Inductive logic programming (ILP) marries Machine Learning and data mining, whose survey can be found in [62, 73]. In particular, Jens Lehmann et al. developed the *DLLearner* [142, 143] to learn $\mathcal{ALC}$ *concept descriptions* from ontologies based on ILP techniques, where the candidate concept descriptions were generated by a downward refinement operator. After that, in [111], they particularly focused on handling larger datasets, such as DBpedia. TBox learning using ILP takes advantages of well-defined refinement operators, which generatively or specifically search towards the target concept. These methods perform quite well when the data quality is relatively high. However, when the dataset suffers from incompleteness (or noise), these methods would drop into local optimum descriptions for concepts due to the incorrect "false" values generated by making CWA.

Association Rule Mining. As a classical data mining method for mining relationships, association rule mining (ARM) is applied in TBox learning problems. Johanna Völker et al. learned $\mathcal{EL}$ axioms from ontologies based on the association rule mining method [247], and in [79], this approach was further extended to learn disjointness axioms. The prototype Goldminer was also implemented. Realising that learning from Semantic Web data suffered from a lack of negative examples when using OWA, Galárraga et al. [84] proposed a *rule* mining model supporting OWA scenario by introducing a new confidence measure in association rule mining. However, these methods mainly use support and confidence thresholds to export the final rules, which work unexpectedly when there is noise or data imbalance. Besides, these methods tend to learn a large number of irrelevant results, which put an extra burden on the end users of ontology learning applications. In addition, association rule mining is also applied to mine rules from dynamic ontologies for providing predictive reasoning.

Statistical Relational Learning. Koller et al. extended DL CLASSIC with nodes in a BN representing probabilistic information of the individuals in a specific class [137], and the model was called P-CLASSIC. It is closely related to the representation in BelNet$^+$. However, in BelNet$^+$, the edges correspond to the specific type of dependency—subsumption. BLP [130] unifies definite logic programs with Bayesian networks. In BLP, ground atoms are mapped to random variables. BelNet$^+$ differs from BLP in that (1) the representation languages are different; (2) concepts are modelled with random variables; (3) schema-level ontology learning is enabled. OntoBayes [256] extends OWL with annotating RDF triples with probabilities and dependencies. All these models have not been applied to TBox learning. In [179], $\mathcal{EL}^{++}$*-LL* was proposed to extend crisp ontological axioms with weights. Using $\mathcal{EL}^{++}$*-LL*, a subset of coherent axioms can be learned from a set of *weighted* $\mathcal{EL}^{++}$ axioms. Besides these works, there are attempts that learn the ABox using graphical models. For example, Rajput and Haider presented a semantic annotation framework that extracts ABox data using Bayesian networks [198].

5.2.8 Conclusion

In this section, we have introduced BelNet$^+$, an approach for ontology learning from *incomplete* information. Such automated ontology construction approaches are interesting tools to deal with the following issues in the context of Knowledge Graph modelling: (1) in Knowledge Graph, making CWA results in noisy data; (2) learning one axiom a time leads to incorrect results in the existence of incompleteness.

In the future, the following aspects could be explored: (1) The current use of exact inference in BelNet$^+$ is not efficient enough for networks with large tree width. It is worth studying this issue and use approximate methods in the future; (2) ABox materialisation on all instances costs too much for a large dataset, such as DBpedia, it might be an idea to find scalable solutions of BelNet$^+$ on very large datasets.

Chapter 6
Understanding Knowledge Graphs

Honghan Wu, Ronald Denaux, Panos Alexopoulos, Yuan Ren and Jeff Z. Pan

Similar to relational databases, knowledge graphs can be utilised to serve in backend of applications by answering structural queries such as SPARQL queries. However, from the users' point of view, what makes Knowledge Graph (KG) unique is that it can serve the end users directly. The enabler of this feature comes from the fact that data semantics are explicitly represented in knowledge graphs instead of being in the business logic layers of applications. But, to realise such feature, the questions are *how such explicit data semantics can be utilised to serve the end users?* and *what kinds of applications knowledge graphs can directly support?* From both academia and industry, many efforts have been put to answer these questions.

In this chapter, we identify and introduce a set of techniques that make knowledge graphs directly available to end users. Among others, we lay special focus on *knowledge graph understanding techniques*, many of which were designed specifically for scenarios in large organisations.

H. Wu (✉)
King's College London, De Crespigny Park, London SE5 8AF, UK
e-mail: honghan.wu@kcl.ac.uk

R. Denaux · P. Alexopoulos
Expert System, Prof. Waksman 10, 28036 Madrid, Spain
e-mail: rdenaux@expertsystem.com

P. Alexopoulos
e-mail: palexopoulos@expertsystem.com

Y. Ren · J.Z. Pan
University of Aberdeen, King's College, Aberdeen AB24 3UE, UK
e-mail: y.ren@abdn.ac.uk

J.Z. Pan
e-mail: jeff.z.pan@abdn.ac.uk

J.Z. Pan et al. (eds.), *Exploiting Linked Data and Knowledge Graphs in Large Organizations*, DOI 10.1007/978-3-319-45654-6_6

Specifically, as summarised in the following list, this chapter will introduce typical applications of understanding a knowledge graph, identify the challenges and present their enabling techniques.

- **Understand Entities** Knowledge graphs are essentially graphs of interlinked entities. The most straightforward application is in helping users understand entities in such graphs. For example, if you search UK in Google, Google's knowledge graph can directly present you the key facts of the United Kingdom like abstract, population and dialling code, and also key related entities of the UK, such as its capital city and main destinations. Section 6.1 discusses the challenges of entity understanding in knowledge graphs with open data and introduces a concept space based approach for summarising individual entities.
- **Exploit Knowledge Graphs** While entity understanding is exciting and very efficient for retrieving common knowledge of entities, sometimes people are more interested in questions for which the answers are not directly available. For example, what are the five coldest years in the UK in the last 50 years. Although such a fact is not directly available (as an attribute of UK), the knowledge graph might have enough data to draw the answer of the question (e.g. through logical based reasoning). In such cases, efficient exploitation techniques might help the user locate the right portion of the knowledge graph and provide them the right facilities to draw the answers by themselves. Section 6.2 introduces a knowledge graph exploitation system that identifies entity description patterns as building blocks for guiding users in their knowledge exploitation tasks.
- **Profile Knowledge Graphs** Data scientists in large organisations might need to make use of multiple knowledge graphs in their daily work, e.g. linking various knowledge graphs to derive knowledge for decision-making. Choosing the right knowledge graphs is the very first step to conduct such tasks. This requires an effective way to determine whether their tasks can be solved by certain knowledge graphs. Knowledge graph profiling is a typical application for serving enterprise users. Section 6.3 presents a knowledge graph profiling approach for users to assess whether knowledge graphs can meet their goals.
- **Reveal the Insights from Knowledge Graphs** Querying knowledge graphs might be one of the most effective and efficient ways to retrieve relevant knowledge from them. However, for end users, the task of constructing correct structural queries itself is already beyond efficient. Automatically generating queries for knowledge graphs or even revealing insights directly from them can be extremely useful techniques for end users to make use of knowledge graphs. Section 6.4 introduces an automatic query generation approach for identifying insights into knowledge graphs.

6.1 Understanding Things in KGs: The Summary of Individual Entities

Understanding entities is one of the typical information needs found in daily searches. For example, to serve users better, major search engines (e.g. Google, Yahoo and Bing) have been generating knowledge cards for popular entities in its search service. If you search *Apple Inc.*, for example, in Google, it will show you a knowledge card like the one in Fig. 3.4 on p. 77.

In this section, we focus on the entity understanding problem in knowledge graphs. Instead of discussing this problem in Web search, we lay special focus on the entity understanding in KGs. In particular, we discuss the problem, point out challenges and describe a summarisation-based approach on RDF data as an illustrating solution. This approach was adopted by a well-known Semantic Web search engine—Falcons (http://ws.nju.edu.cn/falcons/objectsearch/index.jsp) for describing semantic entities in its search results.

In the RDF data model, an entity takes the form of either a URI resource or a blank node. The latter can be viewed as a special case of the URI resource, where an entity is simply not given a unique identification. In the open environment of large organisations, when people are going to share knowledge about the same entities, most likely they will use URIs to denote entities. Hence, in the rest of this subsection, when we mention entities, we mean URI resources.

Using URIs to identify entities makes data integration for the same entities much easier across different knowledge graphs or different portions of the same knowledge graph. This is one of the best advantages to support data integration using Linked Data techniques. However, once the data is integrated, it is not a trivial work to get a quick and comprehensive understanding of the entities in such knowledge graphs. Considering the open data environment in large organisations, there are quite a few challenges in such tasks as follows:

1. **The unordered nature of RDF data**. An RDF graph is essentially a set of triples. This means that triples are orderless. Although this might mean more flexibility for computers to deal with RDF data, such orderless does bring a big obstacle for human users to understand these data. Hence, data organisation approaches are desired to make RDF data easily accessible to human users.
2. **The nature of distributed triples and data redundancies in open data environment**. Unsurprisingly, one entity might be described by many data sources on the Web or in large organisations. The number of data sources can easily rise to tens of thousands if the entity is popular. In addition, redundant data, either duplicated or semantically redundant, might exist in these data sources. Obviously, retrieving these distributed triples and dealing with the redundancies is not an easy task.
3. **The data qualities in the open environment**. While the *Open Data* idea brings a big revolution to the knowledge sharing across various data sources, it also brings another obstacle to understand the entities. When browsing the data about an entity, one often encounters low-quality or inaccurate data, which might have

been introduced by outdated data, bad modelling or even on purpose. Hence, the trustworthiness of data sources is another challenge.

4. **The *huge*[1] data volume challenge**. As mentioned earlier, there could be a large number of data sources describing one entity. This indicates that a large number of triples are relevant to the task of understanding an entity even after the redundancies have been removed. Obviously, a human user cannot easily get a quick and comprehensive understanding when the data space is huge.

In this section, we briefly introduce an approach called *Concept Space based Summarisation* [253], which targets to deal with three of the above four challenges, i.e. numbers 1, 3 and 4. For the challenges of orderless and data quality issues, it proposed a three-dimension organisation, which organises the data in dimensions of concept spaces, RDF triple predicates and data authority. For the huge data volume challenge, it proposed a summarisation technique that assesses the importance of RDF data and extracts the important subset for achieving a quick and comprehensive understanding of entities.

6.1.1 Entity Data Organisation

A widely used approach for organising a large information space is to perform classifications. The basic idea is quite simple, i.e. putting the similar or most relevant parts together and eventually getting a set of classes or groups to organise the information space. Depending on the relations between generated classes or groups, there are different kinds of classification approaches. When the information space is not too large, a set of classes might be good enough. These classes can be disjoint (a data item can only belong to one class) or overlapping (data items can belong to more classes, such as faceted classification[2]). When one-level classification is not good enough, a hierarchical classification might be needed, where the classes are organised into a hierarchical structure, e.g. by applying a multiple step classification. Similar ideas are applied in the Web IR where the search results can be grouped into clusters when a large number of hits are encountered. Generally speaking, there are two challenges to automatic clustering or classification generation: (1) how many clusters or classes are appropriate? and (2) how to generate meaningful clusters/classes?

[1] 'Huge' here means too big to be consumed by human users to get a quick comprehensive understanding.

[2] http://en.wikipedia.org/wiki/Faceted_classification.

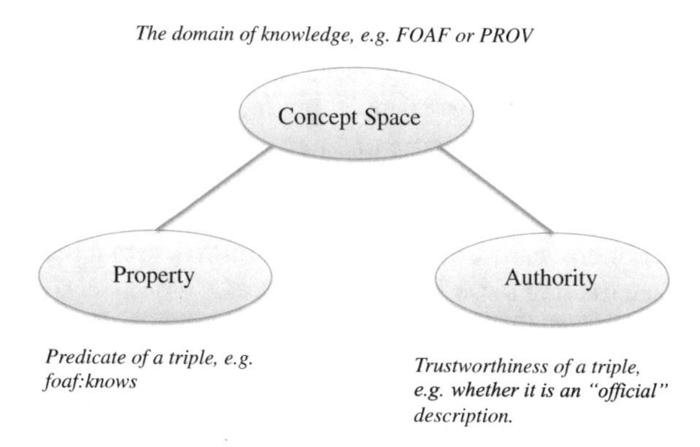

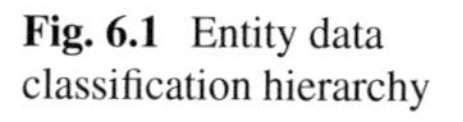
Fig. 6.1 Entity data classification hierarchy

```
Concept Space is a way to group data by application domains
    like FOAF.
```

To tackle the two challenges in an RDF data organisation, the Falcons search engine proposed a hierarchical classification approach (cf. Fig. 6.1) to classify the information space about Semantic Web entities, a.k.a. organising RDF triples that describe entities. As shown in the figure, there are two levels of classification. The concept space level is a disjoint classification (i.e. the domain such as FOAF), while the second level is a faceted classification where a triple is labelled with both a property class (i.e. the predicate such as *foaf:knows*) and an authority class (i.e. the authority assessment result on the data source of the triple). The goal of this approach is to get a classification-based organisation for RDF data which meets the following requirements.

- The classes should be meaningful, which means that they should *make sense to human users*.
- The number of classes should be reasonable in terms of facilitating users' *quick* browsing and understanding.
- There should be indicators about the trustworthiness of the data, so that the users can *filter out* unwanted information in an open environment.

In the first level of classification, RDF triples are classified by a notion called *concept spaces*. The idea of this level of classification is to group data by domains. For example, one person can have information describing her general description (e.g. name, birth date, etc.), social network (e.g. friends of hers) or her profession (e.g. projects she has worked on). Grouping triples by domains can not only give an overall idea of the spheres about the entity, but also help the user in locating interesting information efficiently.

The second level of classification is composed of two facets: (1) Property, (2) Authority. The *Property* facet is to get a fine-grained categorisation within each concept space. For example, putting a person's first name and last name together might make more sense to a human reader rather than having them separated with other attributes (e.g. birth date) in between. The second facet of *Authority* is an indicator as to whether a triple about an entity is from reliable sources. For example, it is more reasonable to believe the birth date of a person from her homepage rather than trusting those from other sources.

The remainder of this section will introduce details about the three classifications of concept space, the property and the authority. In addition, the summarisation approach, which is designed for dealing with the huge volume challenge mentioned before, will be explained.

Concept Space Classification

Compared to entity descriptions in natural language or semi-structured format (e.g. HTML/XML), Linked Data descriptions hold a very good property for information categorisation, which is the use of *vocabularies* in RDF data. A vocabulary is a set of predefined concepts used for describing information. For example, in LinkedMDB knowledge base,[3] movies are given the type of *film*, the *actor* relation is used to specify the acting relations between a movie and its actors, etc. (cf. the movie of Big Daddy[4]). When a data provider generates its linked dataset, there are two options to get suitable vocabularies. One way is to reuse popular vocabularies such as the *FOAF* vocabulary[5] to describe social network information or *geo* vocabulary[6] to specify geographic descriptions. The second approach is to create a new vocabulary. For organising information space about an entity, all the vocabularies in this space form the natural classification by means of domains.

However, the naturally formed vocabulary classes might not be good enough for human users. Different vocabularies might describe information in the same domain. For example, besides *geo* vocabulary mentioned above, there are several other popular vocabularies describing geographic information such as *50K Gazetteer* Vocabulary[7] and *Geographis Ontology*.[8] Obviously, it does not make much sense to separate these information into different classes. The second issue with vocabulary-based classification is that there might be too many concepts to form a reasonable information organisation. Both issues lead to the requirement of a second-level organisation on vocabularies themselves.

---

[3]http://linkedmdb.org/.

[4]http://data.linkedmdb.org/page/film/1004.

[5]http://xmlns.com/foaf/spec/.

[6]http://www.w3.org/2003/01/geo/wgs84_pos#.

[7]http://data.ordnancesurvey.co.uk/ontology/50kGazetteer/.

[8]http://www.telegraphis.net/ontology/geography/geography#.

For the purpose of entity information space classification, the organisation of vocabularies forms the notion of *concept space*. In other words, the *concept space* is a reasonable organisation of a set of vocabularies for the sake of information understanding. Different strategies can be applied to generate concept spaces as follows:

- The first approach is to create concept space by using hierarchical classifications of vocabularies.
- The second approach sometime requires to work on a more fine-grained level, i.e. getting a classification of individual concepts, instead of strictly keeping vocabulary as groups of concepts.
- The third approach can be hybrid which is a combination of the above two approaches.

For example, one hybrid classification can have a class called *general* which contains all general (non-domain-specific) concepts from various vocabularies, e.g. *foaf:firstName* and *schema:gender*, while all other concepts are classified at vocabulary level, i.e. a hierarchical classification of vocabularies.

Regarding the first approach, a.k.a. vocabulary classification, one can use automated approaches by clustering them either using data mining algorithms (e.g. nearest neighbour algorithm [9]) or using more sophisticated techniques from ontology matching community (e.g. considering both structural and literal information). In addition to the automated approach, manual classification can also be applied. Human classification is usually better for understanding but is less scalable. One very useful community contribution of the vocabulary classification comes from the LOV (Live

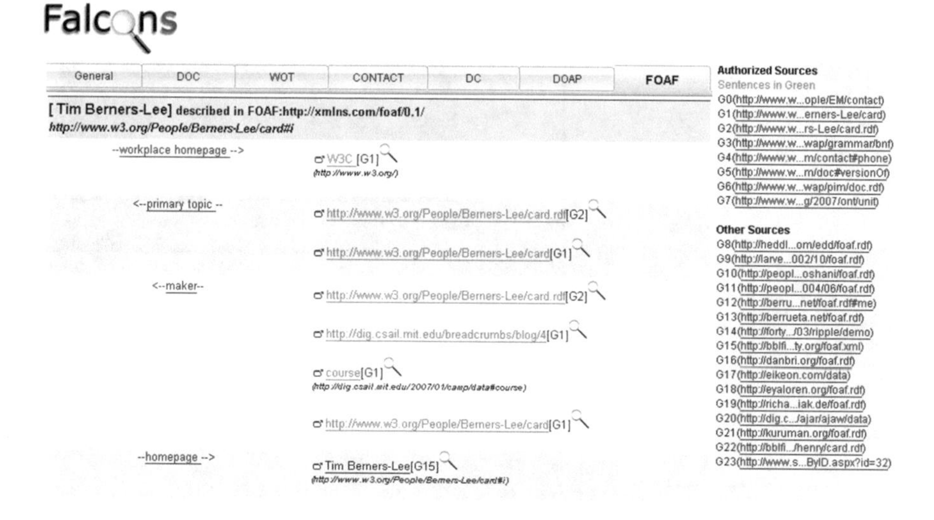

Fig. 6.2 Individual entity summary example: Tim Berners-Lee

One Vision Project) project,[9] where 475 popular Linked Open Data vocabularies are grouped into a hierarchical classification with the first level having 11 groups.

Dealing with heterogeneous RDF data crawled from the Web, Semantic Web search engines are particularly facing the challenges listed on p. 147. To provide a comprehensive entity summary, in the Falcons search engine [49, 253], the above-mentioned hybrid approach is used to create concept space for entity information organisation (see the tabs in Fig. 6.2). The *general* tab is created from the individual concept level, while other tabs are classes of vocabularies. Automated vocabulary organisation is applied in this system. However, it is neither automated classification nor manual classification. A different strategy of selecting top-K is adopted. In particular, of all vocabularies, only the top nine[10] vocabularies are presented in the user interface. The main advantage is that a ranking-based approach is much more efficient than similarity-based classifications. Efficiency is highly valued in on-the-fly summarisation systems like Falcons. More importantly, *based on Falcons datasets (at the time of 2009)*, an analysis result (cf. Fig. 6.3) showed that of all 19,255,731 entities only around 800 entities were described by more than nine vocabularies. This indicates that the ranking-based approach worked well for Falcons 2009 datasets. But things might have been changed after five years of adoption and development of Linked Data knowledge bases. With more datasets published and more linkages created across Linked Open Data cloud, it will not be surprising to see entities being described by many more number of vocabularies. In this case, the classification of vocabularies might be necessary, especially for those popular entities.

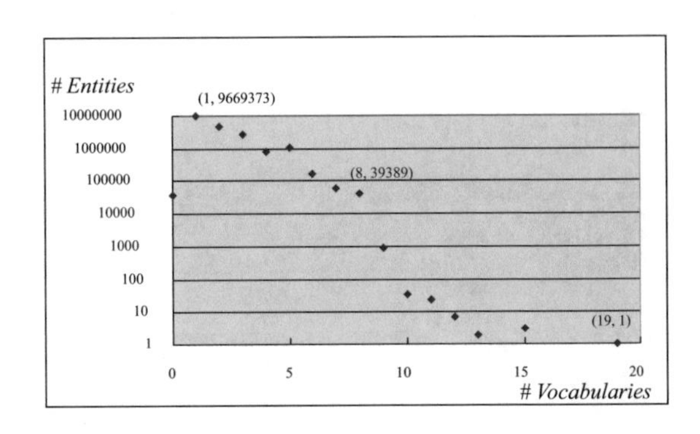

Fig. 6.3 Distribution of #vocabularies of entities

[9] http://lov.okfn.org/dataset/lov/index.html.

[10] Different ranking algorithms can be adopted. In the Falcons system, the ranking is based on the popularity of vocabularies, which is derived from the Falcons 2009 datasets based on document-level frequencies.

Property Facet

The *concept space* organisation groups entity information space by domains. However, in each domain the triples still need to be organised due to the above-mentioned orderless and large volume challenges. Furthermore, when using hierarchical classification, each resulted domain can be either one vocabulary or a hierarchical classification of vocabularies. In the latter case, there is another challenge of (semantical) duplicates which might be annoying to human readers. For example, in the above-mentioned *general* tab, $< me, foaf{:}firstName, Tim >$ *and* $< me, schema{:}firstName, Tim >$ are semantically duplicated information, although different property concepts are used there. We summarise the approaches for two challenges of orderless and duplicates as follows, while leaving the large volume one in the last part of this subsection.

- *Orderless Challenge* The orderless challenge resides in the fact that the RDF model is designed for machine-readable data instead of presenting information for human users. Generally speaking, dealing with this challenge is to find a way to convert the RDF presentation into another format (such as the natural language representation) that is more appropriate for human users to consume. In this sense, natural language generation techniques[11] can be applied here. Alternatively, a less ambitious approach can be to convert RDF data into a semi-structured representation such as an (HTML) list. This is what was implemented in the Falcons entity summary system. The main task is to find a reasonable order to organise triples in a domain. A clustering algorithm is proposed to put similar triples together. The similarity is defined on the predicate of RDF triples, which takes into account the property types (object property or data value property), literal values (URI suffixes and labels) and domain/range similarities (types of domain and value).
- *Duplicate Challenge* In our scenario, duplicate means one piece of information is already stated in or can be effortlessly inferred from another piece. Note that we are talking about human user consumption. Hence, the effortlessly inferred information are those information which can be derived (by most people) from the given information without any noticeable effort. The main idea here is trying to remove "duplicated" information which could annoy (most) human users. Different formal definitions can be given to achieve this goal. In the Falcons system, for a pair of triples $< e, p_1, o_1 >$ and $< e, p_2, o_2 >$, we say the two are "duplicated" *iff* $o_1 = o_2$[12] and $S(p_1, p_2) < C$, where C is a predefined threshold and S is a similarity function. When p_1 and p_2 are from the same vocabulary, $S(p_1, p_2) = 1$, *iff* the vocabulary ontology explicitly specifies $p_1 = p_2$ or $p_1 \sqsubseteq p_2$ or $p_2 \sqsubseteq p_1$; otherwise, $S(p_1, p_2) = 0$. The same approach can be applied on triple pairs where the entity appears in the object position.

---

[11] http://en.wikipedia.org/wiki/Natural_language_generation.

[12] The equality here is a simple implementation, which means string equality for all resource formats including literal values, blank nodes or URIs.

In the Falcons entity summary, the property facet organisation is a set of clusters on de-duplicated properties.

Authority Facet

When the data is coming from various sources, a big problem is whether or to what extent the information from a particular source is correct or trustworthy. In an open environment like the Web, anyone can write anything about an entity. The trustworthiness issue is unavoidable. However, it is subjective that a piece of information is trustworthy or not. There is no one-size-fits-all solution for everyone. An adaptive system for trustworthiness can be a direction one can go for. Sophisticated techniques will be needed for such systems. In the Falcons system, a simple and crisp metric is used for annotating the information, which is the authority facet. Particularly, for a triple $t_1(e) = < e, p_1, o_1 >$, let one of its source document URLs be u, we define a function A. If the entity e has the same domain name of u, $A(t_1(e)) = 1$. The function also holds when the entity appears in the object position of a triple. The idea is that we denote a triple of an entity as authoritative if it is stated in its own domain. As shown in Fig. 6.2, the authoritative RDF triples of Sir Tim Berners-Lee are marked in green. In addition, the user interface (UI) also gives the sources of each assertion, which can be a useful information for the users to determine the trustworthiness of each piece of information.

6.1.2 Summarisation of Entity Data

Even after performing the two-level classification on entity information space, there might still be too many triples for a quick understanding. The number of triples does not have to be too big to pose obstacles for human users' understanding. For example, in the *FOAF* domain, one person might know 50 other people. Listing all these people is obviously not a good presentation. Instead, it is more reasonable to just list the most important friends (e.g. 1 or 2 people) the guy knows. Instead of too many values for one particular aspect, another situation is that there might be too many aspects (properties). Let us say that only 5 triples can be shown for a person's social information but there are dozens of FOAF properties that are used to describe the person. The system needs to determine the most important aspects from among these dozens of properties. Both situations require a summarisation approach to select the most important subset. In the Falcons system, an importance function *Imp* (cf. Formula 6.1) is used to calculate the importance of an RDF triple.

$$Imp(t) = \big(\alpha \cdot CentralityScore(t) + \beta \cdot PrefScore(t)\big) \cdot \sum_{d} DocScore(d) \qquad (6.1)$$

The components of this function are explained as follows.

- *CentralityScore* is a function used to compute the centrality[13] of the triple in the expanded information space of the entity in question. The main idea is to evaluate its importance in the graph structure. The detailed description of the expanded information space generation and the centrality function can be found in [253].
- *PrefScore* is a function to boost the triples which are more informative for human users to understand the entity, such as type assertions that use *rdf:type* as predicates. This function uses a predefined dictionary to record the boosting values for a manually selected list of properties from top vocabularies.
- *DocScore* is a function to calculate the importance of RDF documents. The last component in Formula 6.1 is designed to calculate the sum of importance values of all those documents that contain the triple in question.
- α and β are coefficient values that take the values of 0.35 and 0.65, respectively.

6.1.3 Conclusion

In this section, we discussed the problem of individual entity summarisation in Linked Data knowledge bases. We were assuming that the information space describing an entity is composed of various sources, which can be the case in almost any KG and it is particularly unavoidable in those of large organisations. As we have pointed out, representing knowledge in a Linked Data format facilitates entity summarisation tasks, e.g. URIs make it easier to do integration, and vocabularies form an initial information organisation which is very useful. However, there are still several fundamental challenges in achieving a good entity summarisation. The techniques presented might not work for all cases. But we believe that the problem discussed and the proposed generic framework will benefit those practitioners who need to summarise entities from knowledge graphs. That is why we put more effort in explaining the problems, challenges and possible solutions, while we briefly described the techniques applied in the example application, i.e. the Falcons system. We also omitted the evaluation results of the Falcons entity summary service. Interested readers can get the details from [253].

6.2 Exploring KGs: The Summary of Entity Description Patterns

Individual entity summary helps in a quick understanding of specific entities. This kind of summary is very useful to carry out fact searches, where the user would like to gather simple fact(s) in the domain of interest, such as *who is Barack Obama?* or *what is the current time in New York?*. In such scenarios, the answers to users' questions can be extracted or drawn directly from the knowledge bases (e.g. DBpedia) or

[13]http://en.wikipedia.org/wiki/Centrality.

Knowledge Graph (e.g. Google's KG). However, there are cases for which there are no answers yet and, unfortunately, the (predefined) knowledge inference mechanisms are not smart enough to get meaningful answers. Or perhaps, the users simply would like to exploit the knowledge bases to draw their own conclusions. In such situations, the individual entity summary can hardly help. Instead, users might have to exploit the knowledge base either manually or with the help of some knowledge exploitation tools.

However, exploiting knowledge graphs is definitely not a trivial task, especially when the data volume is large and/or the user is not familiar with the data. Facilities to help users in knowledge exploitation will increase the usability and accessibility of knowledge graphs, which will extend the applications of knowledge graphs into a new type of scenario and probably lead to another killer application just as the individual entity summary has been doing in fact searches.

So far, Linked Data principles and practices are being adopted by an increasing number of data providers, getting as a result a global data space on the Web containing hundreds of LOD datasets [32]. However, the technical prerequisites of using Semantic Web datasets prevent efficient exploitation on these datasets. To tackle this problem, the Linked Data community has been putting a lot of effort into it. We classify such efforts into two groups: (1) work that deals with the extraction of metadata from the datasets and (2) work related to dataset summarisation.

In the following, we briefly explain the related work that extracts metadata from available datasets:

- *LODStats* [66] provides detailed information of LOD datasets such as structure, coverage and coherence, for helping users to reuse, link, revise or query datasets published on the Web. It is a statement-stream-based approach for collecting statistics about datasets described in RDF. *LODStats* presents a smaller memory footprint and better performance and scalability.
- *make-void*[14] is another tool that computes statistics for RDF datasets and generates an RDF output using the VoID vocabulary. It is based on Jena and features advanced criteria, such as the number of links between URI namespaces, or the number of distinct subjects.
- Holst [119] provides an automated structural analysis of RDF data based on SPARQL queries. It identifies, first, a set of 18 measures for structural analysis. Next, it implements the measures in its RDFSynopsis tool, following the two approaches: (1) Specific Query Approach (SQA) and Triple Stream Approach (TSA). Finally, it performs some evaluations over a set of use cases.
- Bohm et al. [42] developed the voiDgen software to analyse RDF graphs and output statistical data in VoID. They propose to compute additional kinds of class, property and link-based partitions, e.g. sets of resources connected via specific predicates. Moreover, they show that distributed analyses of RDF large datasets are feasible by using a distributed algorithm (Map-Reduce).

[14]https://github.com/cygri/make-void.

Considering the requirements of knowledge exploitation, this line of metadata construction work is not effective enough to support exploitation tasks. The main issue is that the generated metadata is too shallow to support comprehensive understanding or knowledge finding.

As regards the works related to dataset summarisation we can say that there are few existing efforts made such as Fokoue et al. [80] for ABox Summary for efficient consistency checking, Zhang et al. [258] for summarising ontologies based on the RDF sentence graphs and Li et al. [145] for a user-driven ontology summarisation. However, both help in the understanding rather than the exploitation, which is usually task oriented.

In the rest of this subsection, we introduce in detail a summarisation-based approach [254] which can not only provide a quick understanding of the dataset in question, but is also able to guide users in exploiting the dataset in various ways.

The main motivation of this summary method is to pursue a *concise* representation or profiles about the knowledge base, which can facilitate knowledge exploitation tasks, e.g. by increasing the efficiency, by revealing hidden knowledge or by providing guidance. We call this summary *Entity Description Pattern*. Such a method has been implemented as an online system, which is available at http://homepages.abdn.ac.uk/honghan.wu/pages/kd.wp3/.

The rest of this subsection is organised as follows: First, we introduce the details of the summarisation definition and generation. Then, we illustrate three exploitation tasks of *(Quick Understanding)* big picture presenting and summary browsing, *(Guided Exploitation)* two query generation methods and *(Dataset Enrichment)* atomic pattern based dataset linkage. Finally, we discuss the graph pattern summarisation techniques and future directions.

6.2.1 What Is the Entity Description Pattern?

Given an RDF graph, the summarisation task is to generate a condensed description which can facilitate data exploitations. Different from the existing ontology summarisation work, we lay special emphasis on identifying a special type of basic graph pattern in the RDF data, which is suitable for data exploitation. The assumption of this special focus is that there exist such building blocks for revealing the constitution of an RDF dataset in a way which can not only help in the understanding of the data but is also capable of guiding RDF data exploitation. The rationale behind the assumption is that RDF data exploitation is usually based on graph patterns, e.g. SPARQL queries are based on the basic graph patterns (BGP).

The main novelty of our summarisation approach is that it summarises an RDF graph by another much smaller graph structure based on atomic graph patterns. The linking structure in such a summary graph can be utilised to significantly decrease search spaces in various data exploitation tasks, e.g. query generation and query answering. Furthermore, statistic results of pattern instances are precomputed and attached to the summary, which can help both in a better understanding of the dataset and more efficient exploitation operations on it.

Specifically, in this section, we propose one definition of such building blocks, i.e. *Entity Description Patterns* (EDPs for short), which is stated in Definition 1.

Definition 1 (*Entity Description Pattern*) Given a resource e in an RDF graph G, the entity description pattern of e is $P_e = \{C, A, R, V\}$, in which C is the set of its classes, A is a set of its data-valued properties, R is the set of its object properties and V is the set of e's inverse properties.

To get a more concise representation of an RDF graph, we define a merge operation on EDPs, which can further condense the graph pattern result (cf. Definition 2).

Definition 2 (*EDP Merge*) Given a set of EDPs $\mathcal{P}$, let $\mathcal{C}$ be the set of all class components in $\mathcal{P}$ and let $G_{\mathcal{P}}(c_i)$ be a subset of $\mathcal{P}$ whose elements share the same class components c_i. Then, *merge* function can be defined as follows:

$$Merge(\mathcal{P}) = \{(c_i, \bigcup_{P_i \in G_{\mathcal{P}(c_i)}} Attr(P_i), \bigcup_{P_i \in G_{\mathcal{P}(c_i)}} Rel(P_i), \bigcup_{P_i \in G_{\mathcal{P}(c_i)}} Rev(P_i)) \mid c_i \in \mathcal{C}\} \tag{6.2}$$

where

- $Attr(P_i)$ denotes the attribute component of P_i;
- $Rel(P_i)$ denotes the relation component of P_i;
- $Rev(P_i)$ denotes the reverse relation component of P_i.

The rationale behind this merge operation is that entities of the same type(s) might be viewed as a set of homogeneous things. Given this idea, we can define an EDP function of an RDF graph as Definition 3.

Definition 3 (*EDP of RDF Graph*) Given an RDF graph G, its EDP function is defined by the following equation:

$$EDP(G) = Merge(\bigcup_{e \in G} P_e) \tag{6.3}$$

EDP Graph The EDP function of an RDF graph results with a set of atomic graph patterns. Most data exploitation tasks can be decomposed into finding more complex graph patterns which can be composed by these EDPs. To this end, it would be more beneficial to know how EDPs are connected to each other in the original RDF graph. Such information can be useful not only for decreasing search spaces (e.g. in query generation) but also for guiding the exploitation (e.g. browsing or linkage). With regard to this consideration, we introduce *RDF data summarisation* as the notion of an EDP graph (cf. Definition 4) for characterising the linking structures in the original RDF graph.

Definition 4 (*EDP Graph*) Given an RDF graph G, its EDP graph is defined as follows:

$$\mathcal{G}_{EDP}(G) = \{< P_i, l, P_j > \mid \exists e_i \in E(P_i), \exists e_j \in E(P_j), < e_i, l, e_j > \in G,$$
$$P_i \in EDP(G), P_j \in EDP(G)\} \tag{6.4}$$

where $E(P_i)$ denotes the instances of EDP P_i. Specifically, if P_i is not a merged EDP, $E(P_i)$ is the set of entities whose EDP is P_i; if P_i is a merged one, $E(P_i) = \cup_{P_k \in P} E(P_k)$, where P is the set of EDPs from which P_i is merged.

Annotated EDP Graph The EDP graphs are further annotated with statistic results. For each node e, it is annotated with a number which is the number of solutions to $Q(x) \leftarrow C_e(x)$. For each edge $l(C_e, C_f)$, there is a tuple of (n_1, n_2), whose elements are the numbers of solutions to $Q_1(x) \leftarrow C_e(x), l(x, y), C_f(y)$ and $Q_2(y) \leftarrow C_e(x), l(x, y), C_f(y)$, respectively.

6.2.2 *How Can the Entity Description Pattern Help in Knowledge Exploitations?*

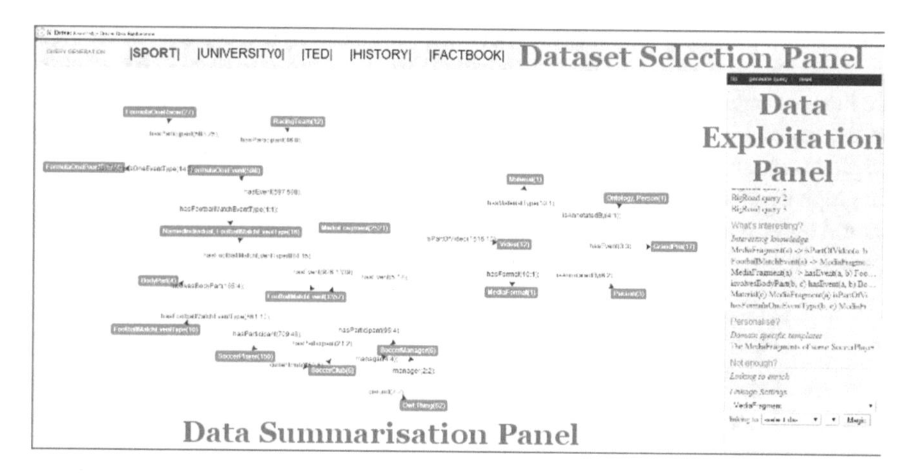

Fig. 6.4 Data exploitation UI

To evaluate and demonstrate the effectiveness of our definition of data building blocks, i.e. EDP, in data exploitation scenarios, we implemented an EDP-based data exploitation system for three types of tasks, i.e. gaining big picture and browsing, generating queries and enriching datasets.

The user interface is shown in Fig. 6.4 which contains three panels. The upper part is the *Dataset Selection Panel*, which displays the list of datasets in the current demo system. To switch to another dataset, one can simply click on its name in this panel. The middle panel is the main interaction and visualisation panel, the *Data Summarisation Panel*. By default, it displays the summarisation of the selected dataset as an interactive graph, i.e. the EDP graph. In other situations, relevant subgraphs of the EDP graph will be shown in the data exploitation process. The right panel is the *Data Exploitation Panel*, which shows a bunch of UI components supporting various data exploitation operations.

Given the UI, we now demonstrate a list of data exploitation scenarios to illustrate how the summarisation can help the data exploitation tasks.

Task 1: The Big Picture and Browsing Operations

When facing an unfamiliar knowledge graph, users usually pursue a quick and rough *big picture* of it before (s)he can assess whether it is interesting or not, e.g. what are the main concepts in this knowledge graph, how are they connected to each other and which are the important parts. To help the users gain answers to these questions quickly, as shown in the *Data Summarisation Panel* of Fig. 6.4, the EDP graph is visualised using force-directed graph drawing techniques.[15] Each node in the graph describes a representative instance. In addition to the node label (concept names), a node is also attached with the number of instances it represents in the dataset. Such statistics (cf. Fig. 6.5) helps to assess the importance of each node in the dataset (in terms of data portions). The relations between (instances of) these nodes are rendered as edges, and such edges are used to calculate groups of closely related nodes, which are in turn rendered as clusters in the graph. Two browsing operations are supported on the summary graph. The first is *node browsing*. By clicking on one node in the graph, users can gain a detailed description of the concept (cf. Fig. 6.5) including the subgraph centralised on this node, which is displayed in the middle panel and the natural language description of the node displayed in a pop-up panel on the left. The second browsing operation is *graph browsing*. After selecting a node, users can keep

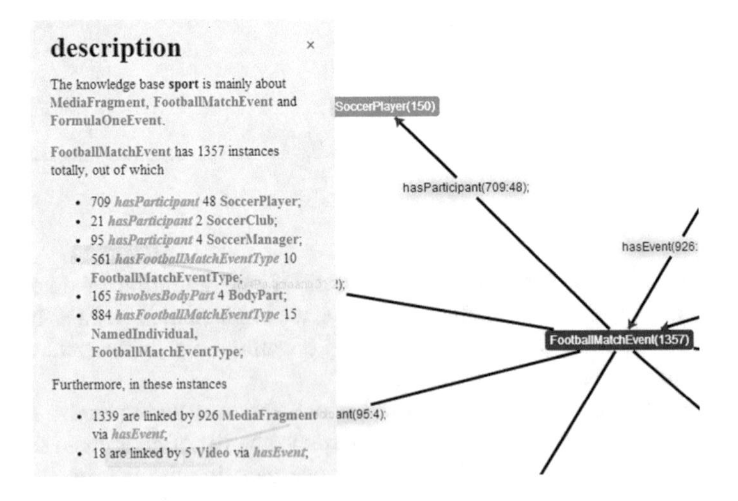

Fig. 6.5 Node browsing

[15] Arbor Javascript Library (http://arborjs.org/introduction) is used for the EDP graph rendering.

selecting/de-selecting interconnected nodes in a current subgraph to grow or shrink it. This operation enables focused investigation on the relations between interested nodes.

Task 2: Query Generation

A typical usage of the knowledge graph is querying it. Query generation techniques [185] are helpful to either a novice or advanced users because the technical skills and dataset knowledge are prerequisites to write SPARQL queries. Based on the EDP summarisation, we implemented two types of query generation techniques. One is called the guided query generation, which generates queries by utilising the EDP graph and statistics information attached in the graph. Such a technique is good at generating queries for revealing the main concepts and relations in knowledge graphs. These two query types are called *Big City Queries* and *Big Road Queries* in the *Data Exploitation Panel* of the system. They are analogous to big cities and highways in a geography map. The other generation technique (which we will discuss in detail in Sect. 6.4) makes use of the links in the summarisation to perform efficient association rule mining. This method is good at revealing insightful knowledge in the data in the form of corresponding graph patterns. Such queries are called *interesting knowledge* in the demo system. Clicking on any of these generated queries will bring out an illustrating subgraph in the middle part of the UI.

Task 3: Knowledge Graph Enrichment

One of the promising features of the Semantic Web techniques is the ability to link data silos to form a more valuable information space. Instead of instance-level linkage or ontology mapping, in our system, we introduce a new data linkage operation on EDPs. Such an EDP-level linkage makes it possible to investigate what kinds of possibilities would be enabled after the cross-knowledge graph EDPs are linked, e.g. previously unanswerable queries might turn out to be answerable by linking another knowledge graph via EDP linkage. In the demo system, users can try to create an EDP linkage between the TED and Factbook datasets and find out how such a linkage can benefit a specific scenario of filtering tenders by country relations.

6.2.3 Conclusion

We described a graph pattern based approach to facilitate knowledge graph exploration. The rationale behind this approach is to provide LEGO-like building blocks for users to play with knowledge graphs. The building block proposed in this section is the so-called EDP, which is a basic entity description graph pattern. We have shown that (i) how can EDPs be constructed as an interactive summary to help in quick understanding; (ii) how can EDPs be utilised to generate *interesting* queries under different interestingness definitions; (iii) how can EDPs from different sources be linked to each other so that unanswerable queries are turned to be answerable after this data enrichment. The future work will focus on investigating the properties of the summary and in-depth studies in the above scenarios.

6.3 Profiling KGs: A Goal-Driven Summarisation

6.3.1 Motivating Scenario and Problem Definition

The knowledge graph (KG) summarisation approach we described in the previous section treats the summarisation task in an application and user-independent way by producing generic summaries whose usefulness is limited to an all-purpose very high-level overview of the data. By contrast, in this section, we are interested in facilitating the generation of requirement-oriented and task-specific KG summaries that may help knowledge engineers and data practitioners assess whether and to what extent a given KG is suitable for the task at hand.

To that end, in this section, we describe a goal-driven KG summarisation framework that may be used to examine and evaluate the suitability of knowledge graphs for (re-)use in particular application domains and scenarios. Within this framework users are able to define and execute custom summarisation processes to generate useful KG summaries. A custom summarisation process can be seen as an orchestration of primitive predefined parameterisable graph analysis processes, each of which may deal with a different aspect of the graph. More importantly, such a process is linked to a particular goal/problem/need that it is supposed to serve, thus forming a reusable knowledge component that can be shared among multiple users with similar needs.

6.3.2 Framework Description

The proposed summarisation framework aims to enable its intended users to answer the following question: *"Given an application scenario where a knowledge graph is required, how suitable is a given existing graph for the purposes of this scenario?"*. To answer this question, users normally need to be able to: (i) explicitly express the requirements that a KG needs to satisfy for a given task or goal and (ii) automatically measure/assess the extent to which a KG satisfies each of these requirements and compile a summary report.

To implement these two capabilities, we follow a *checklist-based* approach. Checklists are practically lists of action items arranged in a systematic manner that allow users to record the completion of each of them and they are widely applied across multiple industries, like healthcare or aviation, to ensure reliable and consistent execution of complex operations [102]. In our case, we apply checklists to define and execute custom KG summarisation tasks in the form of lists of goal-specific requirements and associated summarisation processes. In the following paragraphs, we explain how such tasks and processes may be represented, created and used.

Summarisation Task Representation

To represent custom summarisation tasks according to the aforementioned checklist paradigm, we adopt the *Minim model* [260] that allows us to represent for concrete instances of summarisation tasks the following information:

- The **Goals** the KG summarisation task is designed to serve. In the Minim's terminology [260], they are called constraints and are used to denote the purpose of the summarisation task and the intended use of the produced summary. This is important as different tasks may have different purposes (e.g. the requirements for checking whether a KG is appropriate for disambiguation may be different from those required for question answering) and, thus, the goal-related information is crucial for selecting an already defined task in a given application scenario.
- The **Requirements** (or checklist entries) against which the summarisation task evaluates the KG. For example, we may wish to assess whether a KG contains particular information about a given domain or topic or that it satisfies particular quality criteria (e.g. consistency). The number and nature of the requirements depend practically on the goal of the summarisation task and thus may be substantially different among different application scenarios.
- The **Data Analysis Operations** that the summarisation task employs in order to assess the satisfaction of its requirements. In the Minim's terminology, these operations are called rules and practically they take many forms, from simple execution of queries to complex data processing and analysis algorithms like graph analysis or topic modelling. The assessment of a given requirement may require the execution of multiple operations while the same operation may be used to assess multiple requirements.

Summarisation Task Creation

To create a summarisation task one needs to define its goal(s), its requirements and the association to these operations. Some high-level requirements that we have already identified and may be used for multiple goals are the following:

- **Evaluate the KG's coverage of a particular domain/topic**: This requirement aims to measure the extent to which a KG describes a given domain or topic. This can be at schema level (e.g. how many and which concepts or relations are defined), at instance level (e.g. how many and which instances of a given concept or relation does the dataset have) or with more complex operations (e.g. comparison with a corpus).

- **Evaluate the KG's labelling adequacy and richness**: This requirement aims to measure the extent to which the KG's elements (concepts, instances, relations, etc.) are accompanied by representative and comprehensible labels in one or more languages. This can be useful to assess two factors: (i) the comprehensibility of the KG, i.e. the ease with which human consumers can understand and utilise the KG's data and (ii) the quality and usefulness of a KG as a thesaurus term.
- **Evaluate Connectivity**: This requirement checks the existence of paths between concepts or entities, i.e. whether it is possible to go from a given concept to another on the graph and in what ways. This is an important aspect of a KG related, for example, to its ability to answer queries involving particular related entities.
- **Evaluate Ambiguity**: This requirement aims to measure the extent to which entities in the KG share common identifiers and labels and thus are prone to miscomprehension within an application scenario.

Each of the above requirements can be implemented by means of one or more data analysis operations. Some operations which we have already defined within the framework are the following:

- Check the existence of a particular element (concept, relation, attribute, instance, axiom) in the dataset or that of a relational path between particular concepts or instances.
- Measure the number of ambiguous entities in the KG.
- Measure the number of labelled entities.

Summary Generation

For the generation of goal-specific KG summaries, we have developed a tool that may take as input one or more KGs and a summary goal and run on them specified summarisation tasks that correspond to this goal. The output of this tool is a detailed report about the input KGs, describing whether and to what extent do they satisfy each requirement. The next section provides a concrete example of this output in the context of an actual use case where we applied the framework.

6.3.3 Implementation

In this section, we describe the architecture of the KG summarisation system, in terms of the components it comprises and the main functions that these components serve. The architecture is depicted in Fig. 6.6 and consists of three layers, namely the **data layer** where all the necessary data are stored, the **application layer** that implements the requirement functionalities for the generation and management of summaries and, finally, the **presentation layer** that implements the system's user interface.

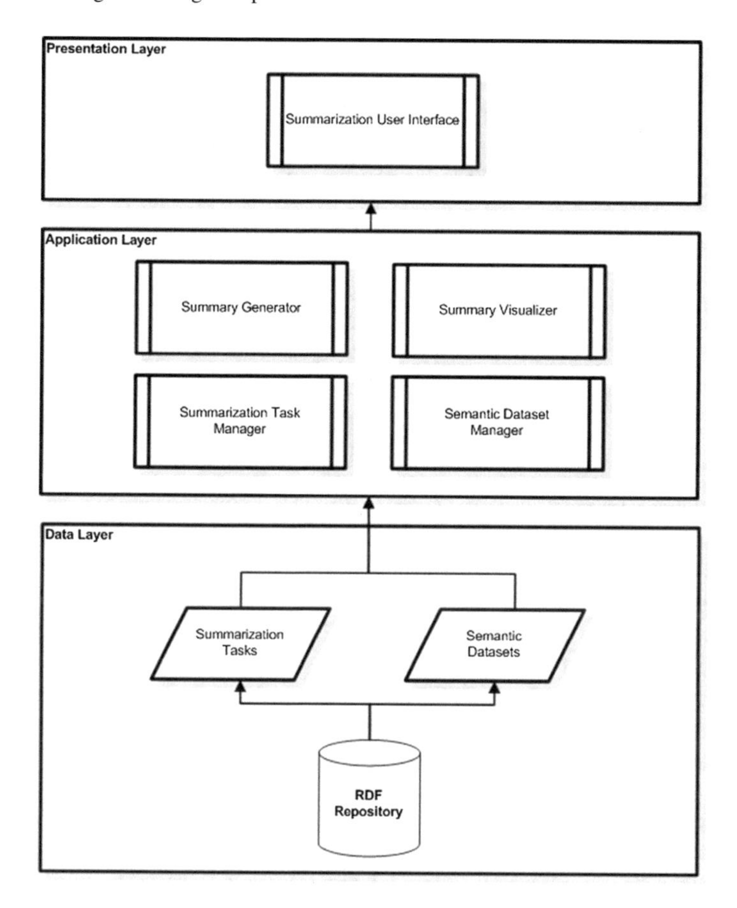

Fig. 6.6 Semantic data summarisation platform architecture

In more detail, the main components of the system are the following:

RDF Repository This is a repository that stores two types of data, namely (i) the KGs that the system is supposed to analyse and produce relevant summaries about and (ii) the summarisation tasks that are to be applied to the KGs used for the summary generation. For performance and reliability reasons, the KGs are stored locally and updated periodically from their original sources. As for the summarisation tasks, these KGs are defined and stored as instances of the ontological schema of Fig. 6.7.

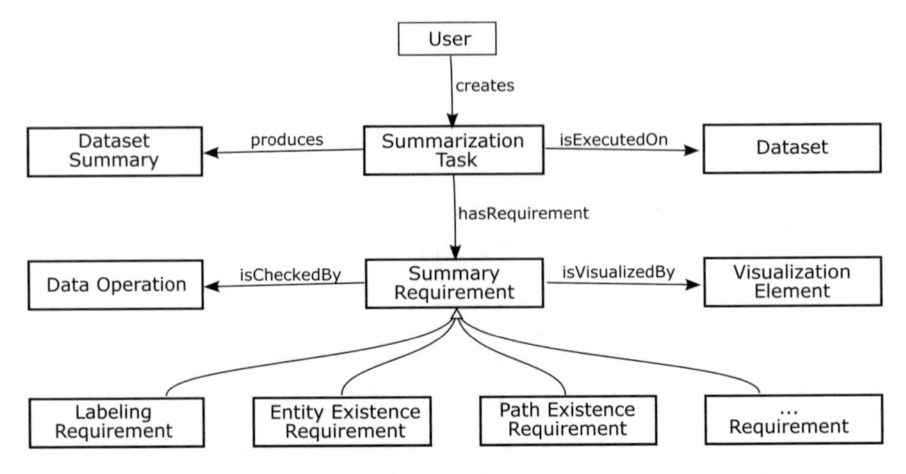

Fig. 6.7 Ontological schema for summarisation tasks

Knowledge Graph Manager This is an API for accessing and manipulating, at a low level, the KGs that are to be used within the system and it is primarily used by the Summary Generator component.

Summarisation Task Manager This component provides a comprehensive API for accessing and manipulating, at a low level, the summarisation tasks that users define and use. It interacts with the RDF repository where the tasks are stored as well as the components that require its information like the Summary Generator and the Summary Visualiser.

Summary Generator This is the core component of the overall system, primarily responsible for executing the user-defined summarisation tasks and producing corresponding summaries that are also stored in the system's repository. It implements a predefined set of data operations that the user may define in his/her summarisation tasks and executes them when invoked.

Summary Visualiser This component produces, for a given generated summary, a set of visualisation elements that are to be shown to the user at the Presentation layer. These elements are typically predefined and associated to summarisation requirements, as shown in Fig. 6.7. This association is based on an a priori analysis about what visualisation elements best visually communicate the information a requirement generates.

Summarisation User Interface This is the main channel through which the system's users are able to define summarisation tasks and produce relevant summaries. The task definition is facilitated primarily by means of forms while the produced summaries are visualised according to the dashboard paradigm. Figure 6.8 illustrates the main elements of the user interface.

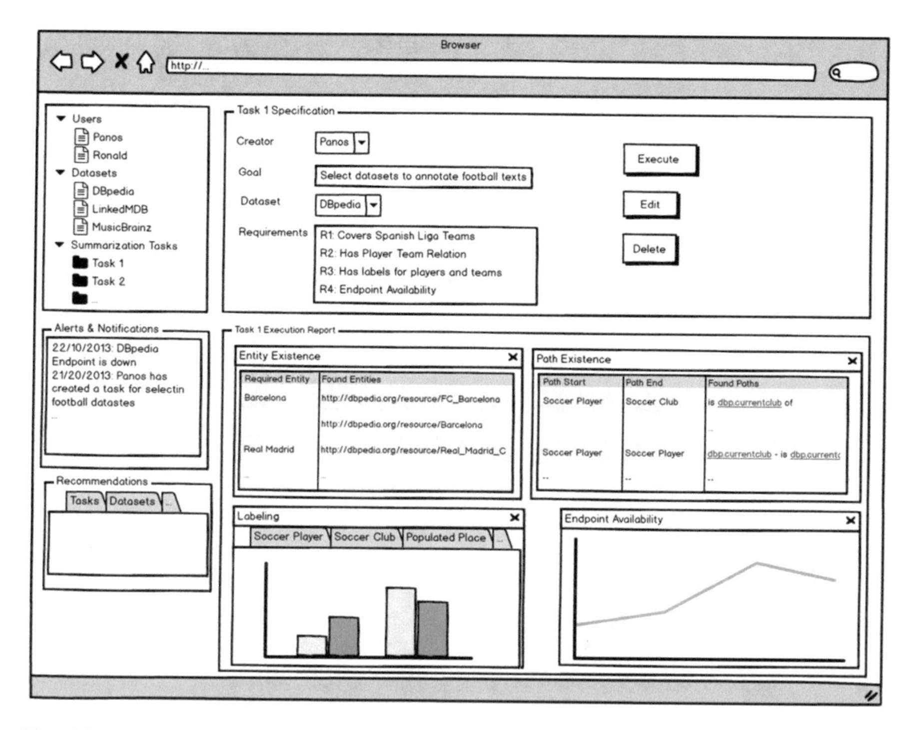

Fig. 6.8 Summarisation system user interface

6.3.4 *Application Example*

In this section, we go through a specific case of how the goal-driven summarisation can be used in practice. The person interested in finding some knowledge graph works for a media company which manages a large set of images with textual descriptions about the Spanish Liga. The company wants to make these images easier to find using a semantic search over the textual descriptions. One of the prerequisites for a semantic search is the availability of a knowledge graph which describes the domain accurately. Hence, the company needs a knowledge graph about the Spanish Liga. Using the goal-driven summarisation platform, the company can monitor public knowledge graphs to identify the existing KGs that could be suitable for their goal.

To start the definition of their goal, the user will be presented with the form shown in Fig. 6.9, where they fill out "Annotate texts about the Spanish Football League" to describe their **goal**.

Next, they start defining their **requirements**, using a form such as that shown in Fig. 6.10, where they can describe their requirement via text, choose between various available **dataset operations** and fill out specific values relevant to their requirement. For example, in this case they specify the requirement to "mention all the teams in the

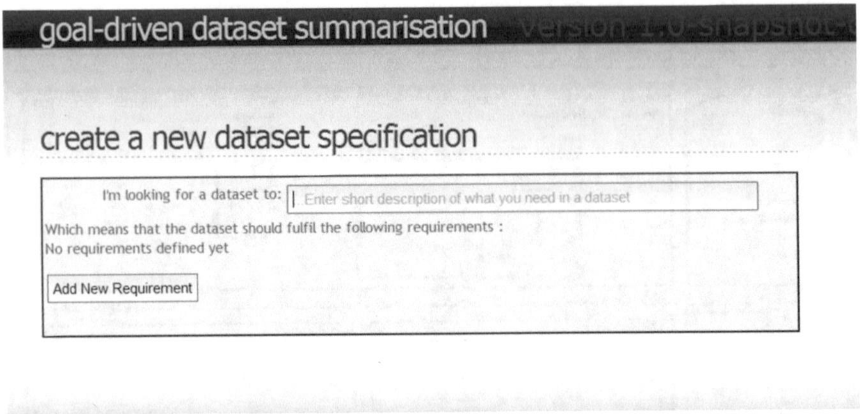

Fig. 6.9 The initial form that users can use to describe their goal (for what they want a knowledge graph) and to start adding requirements to fulfil that goal

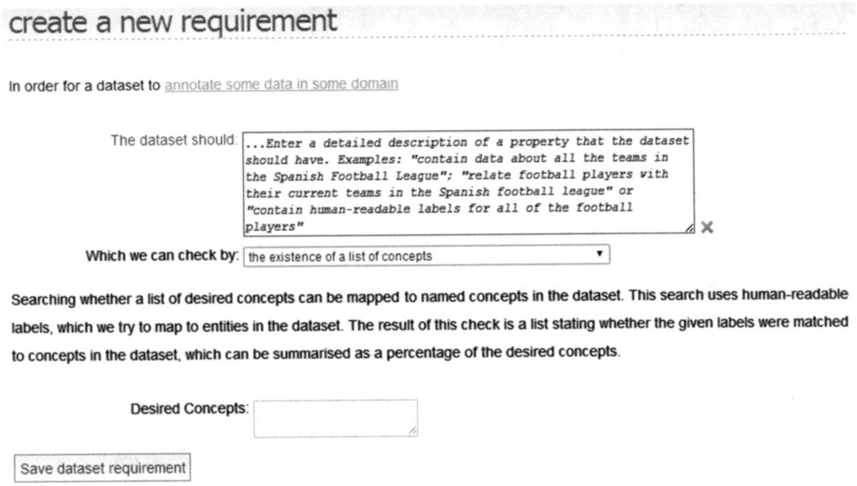

Fig. 6.10 Form for entering a new requirement to be associated with a summarisation goal. In this case, the user can define a list of entities that must be defined in the desired knowledge graph

Spanish Liga" and enter the list of teams they know (e.g. Real Madrid, FC Barcelona). They continue doing this for more requirements they identify.

Once they are happy with their goal and requirements' definition, the platform schedules the summarisation task and analyses known public knowledge graphs in order to determine their suitability for this particular goal. The results are then presented to the user as shown in Fig. 6.11 as a **checklist** for the various known knowledge graphs, which can then be inspected in more detail using a **dashboard** as shown in Fig. 6.12.

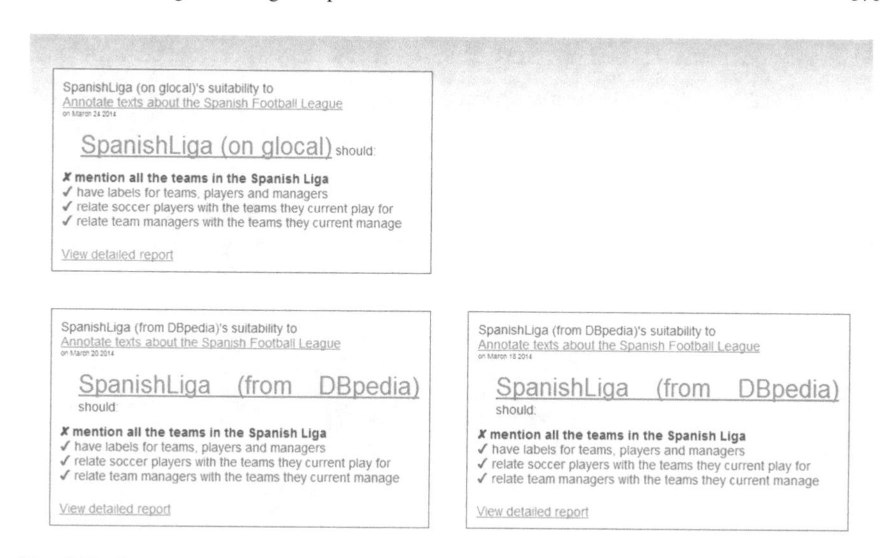

Fig. 6.11 Segment of a *Recent Dataset Summaries* page showing the checklist reports for some datasets

6.3.5 Conclusion

In this section, we presented a recent work on enabling end users to define relevant goals related to the task of summarising knowledge graphs in terms of such precise goals. Such a platform can be used to find suitable publicly available knowledge graphs and is a potentially good complement to general-purpose summaries of entities and knowledge graphs. While general-purpose summarisations help users to understand what type of information is contained within a knowledge graph, goal-driven summarisations help users to determine the suitability of a knowledge graph for a particular purpose.

6.4 Revealing Insights from KGs: A Query Generation Approach*

To explore knowledge graphs, semantic queries, i.e. graph patterns with variables, are important means for users to retrieve matching results. Standard query languages have been developed and supported with tools. Semantic data can be exploited by queries [215], such as those in the standard RDF query language SPARQL.

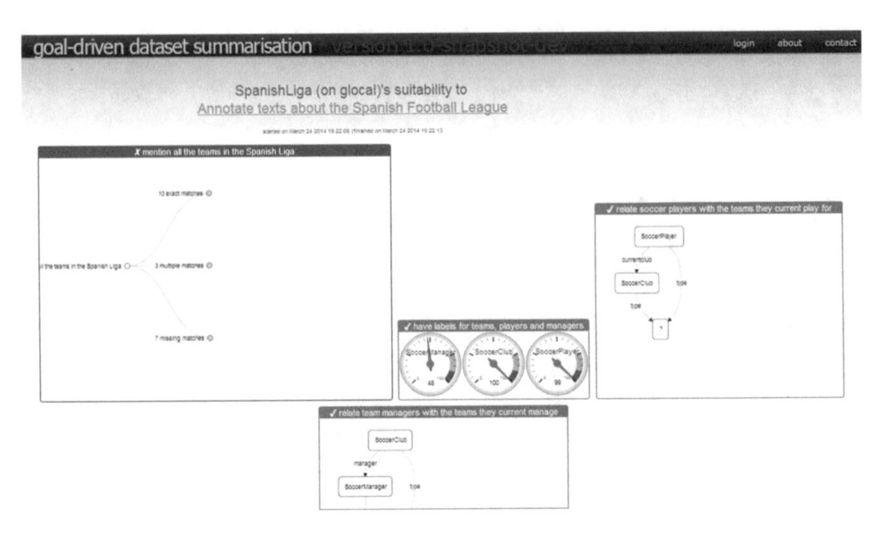

Fig. 6.12 Dashboard report summarising how suitable a dataset (the SpanishLiga dataset) is for a given specification

However, novice users tend not to be familiar with RDF and SPARQL. Also, the users are likely to be unfamiliar with external datasets that are linked to their local ones. Given some target dataset(s), it would be desirable to have a service to recommend insightful queries to users. For example, in the Lehigh University Benchmark (LUBM),[16] the following queries Q1 and Q2 have the same results under RDF simple interpretation (i.e. without reasoning [81, 110]).

```
#Q1: Return those who take a Course
SELECT ?x WHERE {
  ?x lubm:takeCourse ?y .
  ?y rdf:type lubm:Course . }
#Q2: Return Undergraduates who take a Course
SELECT ?y WHERE {
  ?x lubm:takeCourse ?y .
  ?x rdf:type lubm:Undergraduate .
  ?y rdf:type lubm:Course . }
```

This implies that only undergraduate students are taking courses in the dataset, which might be somehow *surprising*, as postgraduate students are *not* taking any courses. This gives users a sense of quality of the query answers when reasoning is disabled.

Query generation (QG) has been studied in relation to database (e.g. [217]) and actual data (e.g. [162]), with the main motivation to generate queries for testing databases. A related research problem is query+ recommendation (QR), where query logs are widely used to generate queries based on the querying and browsing behaviours of users [48, 259]. Similar to QG for testing databases, the work on QG for

[16]http://swat.cse.lehigh.edu/projects/lubm/.

testing Semantic Web engines [92, 93], based on ontologies or parameterisations, is available. d'Aquin and Motta [64] proposed a QG approach based on the formal concept analysis, which uses a computationally complex ontological reasoning. In this section, we propose a tractable query generation approach based on data summarisation and graph patterns.

Preliminaries

We assume the readers to have basic knowledge about RDF and SPARQL. An RDF graph G can be divided into mutually disjoint schema graph G_s and instance graph G_i. An RDF graph is an instance graph if it only contains *type triples* of form $< x, rdf{:}type, A >$ or relation triples $< y, R, z >$, where A is a class, R is a user-defined property and x, y, z are resources. With these notions, to facilitate query generation, we define a graph as follows:

Definition 5 (*Graph*) A *labelled, directed multiple graph* (graph for short) $G = \langle N, E, M, L \rangle$ is a 4-tuple, where N is the set of nodes, E is the set of edges, $M : E \rightarrow N \times N$ maps an edge to an ordered pair of nodes, L is the labelling function that for each node $n \in N$, its label $L(n)$ is a set of URI references, and for each edge $e \in E$, its label $L(e)$ is a URI reference.

For any RDF instance graph $\mathcal{D}$, let T be its set of types and TT be its set of type triples, a unique graph $G_{\mathcal{D}} = \langle N, E, M, L \rangle$ can be constructed as follows, where type triples are aggregated as the labels of nodes, and relation triples are represented as directed, labelled edges in the graph: (i) $N = \{x| < x, p, o > \in \mathcal{D}$ or $< s, p, x > \in \mathcal{D}\} \setminus T$; (ii) $E = \mathcal{D} \setminus TT$; (iii) $\forall < s, p, o > \in E, M(< s, p, o >) = (s, o), L(< s, p, o >) = p$; (iv) $\forall n \in N, L(n) = \{C| < n, rdf{:}type, C > \in \mathcal{D}\}$.

Definition 6 (*Graph Operations*) A graph $\langle N_1, E_1, M_1, L_1 \rangle$ is a *subgraph* of another graph $\langle N_2, E_2, M_2, L_2 \rangle$ IFF $N_1 \subseteq N_2$, $E_1 \subseteq E_2$, $\forall e \in E_1$, $M_1(e) = M_2(e)$, $L_1(e) = L_2(e)$ and $\forall n \in N_1, L_1(n) \subseteq L_2(n)$.

A graph G is the *intersection* of graphs G_1 and G_2 IFF it is the largest subgraph of both G_1 and G_2.

Two graphs $\langle N_1, E_1, M_1, L_1 \rangle$ and $\langle N_2, E_2, M_2, L_2 \rangle$ have a *union* IFF $\forall e \in E_1 \cap E_2$, $M_1(e) = M_2(e)$, $L_1(e) = L_2(e)$. Their *union* is a graph $\langle N, E, M, L \rangle$ such that $N = N_1 \cup N_2$, $E = E_1 \cup E_2$, $\forall e \in E_1 \cup E_2$, $M(e) = M_1(e)$ or $M(e) = M_2(e)$, $L(e) = L_1(e)$ or $L(e) = L_2(e)$, and $\forall n \in N_1 \cup N_2$, $L(n) = L_1(n) \cup L_2(n)$.

SPARQL supports many different patterns. In this section, we are interested in studying the basic graph pattern (BGP) and the `FILTER NOT EXISTS`. A BGP is a set of triples with variables. A solution to a BGP is a mapping of the variables to resources or blank nodes in the RDF graph, such that the mapped graph is a subset of the RDF graph. For a BGP `bgp`, the solution to `FILTER NOT EXISTS bgp` are the mappings of variables to resources or blank nodes, such that the mapped triples do not occur in the RDF graph. We also slightly abuse the notion to say that, two queries Q_1 and Q_2 can be combined into a composite query, denoted by $Q_1 \wedge Q_2$, by including the variables and triples in both Q_1 and Q_2.

In this section, we are interested in *counjunctive queries without non-distinguished variables* and their complements and composites. A conjunctive query BGP contains only the type triples and the relation triples, where only the subject of type triples, and subject and object of relation triples can be variables. Queries with non-distinguished variables are supported by the new SPARQL standard, where all variables in queries must be bounded to named entities in the RDF graph. With the above considerations, BGP queries discussed in this section can be regarded as a special kind of graph.

Definition 7 (*Graph Pattern*) A *graph pattern* is a graph in which some nodes are variables.

For any conjunctive query BGP, a unique graph pattern can be constructed, and vice versa. We use $Q(GP)$ to denote a query constructed from a graph pattern GP. It is obvious that $Q(GP_1) \wedge Q(GP_2) = Q(GP)$, where GP is the union of GP_1 and GP_2.

Query answering can also be realised by graph pattern matching:

Definition 8 (*Instance*) A *substitution* $\mathbf{S} = (\upsilon_1 \rightarrow \upsilon_2)$ replaces a vector of variables υ_1 with a vector of URI references/literals/blank nodes/variables υ_2. A substitution is variable-free IFF υ_2 contains no variable. A graph G is an *instance* of a graph pattern G', denoted by $G : G'$, IFF there exists a substitution $\mathbf{S}$ such that $G'_S = G$. Given a dataset $\mathcal{D}$, we use $I_{\mathcal{D}}(G')$ to denote the set of all subgraphs of $\mathcal{D}$ that are instances of G'. Obviously, $G \in I_{\mathcal{D}}(G)$ since an empty substitution $(\emptyset \rightarrow \emptyset)$ exists. And $I_{\mathcal{D}}(G) = \{G\}$ when G contains no variable. A graph G is a υ-*instance* of a graph G' w.r.t. $\mathcal{D}$ and a vector of variables υ IFF there is a variable-free substitution $(\upsilon \rightarrow \upsilon')$ such that $G'_{(\upsilon \rightarrow \upsilon')} = G$ and $I_{\mathcal{D}}(G) \subseteq I_{\mathcal{D}}(G')$. We use $I_{\mathcal{D},\upsilon}(G)$ to denote the smallest set of all υ-instances of G w.r.t. $\mathcal{D}$ and υ.

When the $\mathcal{D}$ is clear from context, we omit it in the notations.

The following proposition shows the relation between a solution of a query and an instance of the graph pattern of the query:

Proposition 1 (Query Answering) *Given a dataset $\mathcal{D}$, for any conjunctive query BGP Q of a vector of variables υ_1, let GP be its corresponding graph pattern, then if the mapping from υ_1 to υ_2, a vector of resources, is a solution of Q w.r.t. $\mathcal{D}$, then $GP_{(\upsilon_1 \rightarrow \upsilon_2)} \in I(GP)$. And if $G \in I(GP)$ with substitution $(\upsilon_1 \rightarrow \upsilon_2)$, then the mapping from υ_1 to υ_2 is a solution of Q w.r.t. $\mathcal{D}$.*

With the above notations, we can investigate the query generation problem by observing graph patterns in datasets.

6.4.1 Candidate Insightful Queries

Given a target graph $\mathcal{D}$, the generation of *candidate insightful queries* for $\mathcal{D}$ can be regarded as a process of identifying typical graph patterns (or typical graph pattern pairs) having instances within $\mathcal{D}$, such that these typical graph patterns (or typical graph pattern pairs) provide users with some insights into the structure of $\mathcal{D}$.

Typical Graph Patterns

Typical graph patterns are concerned with the structured relations among domain objects. While schema graphs (or ontologies) specify some global structure, typical instance graph patterns relate to some possible additional structure in the current version of the graph. There can be different kinds of typical graph patterns, such as star-shaped graphs, shallow tree shaped graphs, deep tree shaped graphs and graphs with loops.

Definition 9 (*Looped Graph Pattern*) A graph pattern is a *looped graph pattern* if it contains a circle of nodes.

A looped graph pattern is a *variable-looped graph pattern* if it contains a circle of variables.

A query is a *(variable-)looped query* if its corresponding graph pattern is a (variable-)looped graph pattern.

In this section, we are particularly interested in graphs with a looped graph pattern, since loops reveal the multiplicity of the connections between objects, i.e. objects are connected in the dataset via multiple paths. Furthermore, nominal-free[17] ontologies are not sufficiently expressive to accurately represent loops. In other words, graphs with loops might give users some insights into complex relations among objects that are not captured in the corresponding ontology.

Next, we first introduce the notion of graph pattern correspondence and then revisit looped queries.

Graph Pattern Correspondence

Graph pattern correspondence is concerned with the relationships between two groups of objects. In the Introduction, we gave some examples of queries with the same set of answers. They can be formally described by the correspondence of two graph patterns as defined below:

Definition 10 (*Graph Pattern Correspondence*) Given an RDF instance graph $\mathcal{D}$, two graph patterns GP_1 and GP_2 correspond on a vector of variables υ IFF there is a variable-free substitution $\upsilon \rightarrow \upsilon'$ such that $GP_{1(\upsilon\rightarrow\upsilon')} \in I_{\upsilon,\mathcal{D}}(GP_1)$ and $GP_{2(\upsilon\rightarrow\upsilon')} \in I_{\upsilon,\mathcal{D}}(GP_2)$. υ' is called the υ-correspondence of GP_1 and GP_2 w.r.t. $\mathcal{D}$.

We use $C_{\mathcal{D},\upsilon}(GP_1, GP_2) = \{\upsilon' | \upsilon'$ is a υ-correspondence of GP_1 and GP_2 w.r.t. $\mathcal{D}\}$ to denote the set of all υ-correspondences. In the rest of the paper, we omit $\mathcal{D}$ from the notations when it is clear from context.

From the above definition, it is obvious that two graph patterns GP_1 and GP_2 correspond on a vector of variables υ IFF there is a solution to $Q(GP_1)$ and a solution to $Q(GP_2)$ that have the same value assigned to υ. While $C_{\mathcal{D},\upsilon}(GP_1, GP_2)$ actually indicates the different values of υ that can be shared by solutions of $Q(GP_1)$ and $Q(GP_2)$. The following theorem shows the relation between graph pattern correspondence and conjunctive query answering:

[17]Nominal is one of the OWL features that could introduce scalability problem for reasoning.

Theorem 1 *Two graph patterns GP_1 and GP_2 correspond on variables υ IFF $Q(GP_1) \wedge Q(GP_2)$ has a solution.*

$C_{\mathcal{D},\upsilon}(GP_1, GP_2) = \{\upsilon' \mid \upsilon'$ *is the value assigned to* υ *in some solution of* $Q(GP_1) \wedge Q(GP_2)$ *w.r.t.* $\mathcal{D}\}$.

This result can be utilised to generate the following kinds of insightful queries.

1. **Queries with strong Correspondence**: For two graph patterns GP_1 and GP_2, let υ be a vector of all their shared variables, $Q(GP_1)$ and $Q(GP_2)$ are insightful queries with strong correspondence if $\mid C_\upsilon(GP_1, GP_2) \mid$ is higher or lower enough w.r.t. $\mid I_\upsilon(GP_1) \mid$ or $\mid I_\upsilon(GP_2) \mid$, which indicates that $Q(GP_1)$ and $Q(GP_2)$ share a lot, or there are very few solutions on variables in υ, respectively.
   This is because, $\mid I_\upsilon(GP_1) \mid$ and $\mid I_\upsilon(GP_2) \mid$ are the numbers of different solutions assigned to υ in $Q(GP_1)$ and $Q(GP_2)$, respectively, and $\mid C_\upsilon(GP_1, GP_2) \mid$ is the number of different solutions assigned to υ in both $Q(GP_1)$ and $Q(GP_2)$. When $\frac{\mid C_\upsilon(GP_1,GP_2) \mid}{I_\upsilon(GP_1)}$ is close to 1, it indicates that most of the solutions to $Q(GP_1)$ can also be regarded as solutions to $Q(GP_2)$ for all variables they share. When it is close to 0, it indicates that only very few solutions to $Q(GP_1)$ can be regarded as solutions to $Q(GP_2)$. Both queries with high shared solutions and queries with low shared solutions are treated as insightful queries.
   Note that such a solution-sharing relation is not symmetric, i.e. it is possible that $Q(GP_1)$ shares many solutions with $Q(GP_2)$ but $Q(GP_2)$ only shares a few solutions with $Q(GP_1)$.
2. **Queries on Exceptions**: With the correspondence defined in Definition 10 we can generate insightful queries due to exceptions.

   - For a pair of queries Q_1 and Q_2 with very high correspondence, $Q_1 \wedge \{$`FILTER NOT EXISTS` $(Q_2)\}$ (or $Q_2 \wedge \{$`FILTER NOT EXISTS` $(Q_1)\}$) will be an insightful query on exceptions. Obviously, if two queries share a lot of solutions, then the solutions that do not belong to both of them are quite interesting to users.
   - For a pair of queries Q_1 and Q_2 with strong low correspondence, $Q_1 \wedge Q_2$ will also be an insightful query on exceptions. If two queries share very few solutions, then the solutions that belong to both of them are interesting to users.

An extreme case of queries under exception is empty queries. An empty query is a query that does not have any solution on the given RDF dataset, as opposed to arbitrary datasets [15], where a query is empty if it does not have any solution for any dataset. Our notion of emptiness is based on the input instance graph(s), while the arbitrary datasets' notion of emptiness is forced by the input ontology (schema graph). Our notion of empty queries is weaker and cannot be checked by their approach.

Now that we introduce the notion of graph pattern correspondence, let us revisit looped queries, some of which can be regarded as a special extension of queries with high correspondence: for two graph patterns GP_1 and GP_2, if there is a path of variables $\upsilon_1, \upsilon_2, \ldots, \upsilon_n$ in GP_1 and a path of variables $u_1, u_2, \ldots, u_m$ in GP_2, $\upsilon_1 = u_1$ and $\upsilon_n = u_m$ are variables shared by GP_1 and GP_2 and GP_1 and GP_2 have high correspondence w.r.t. vector $\langle \upsilon_1, \upsilon_n \rangle$, then $Q(GP_1) \wedge Q(GP_2)$ is a looped query.

This is obvious, since by combining GP_1 and GP_2 we have a looped graph pattern containing variable loop $v_1, v_2, \ldots, v_n, u_{m-1}, \ldots, u_2, v_1$. And this looped graph pattern can be transformed to $Q(GP_1) \wedge Q(GP_2)$. If GP_1 and GP_2 have high correspondence, it indicates that the solutions to $Q(GP_1)$ construct loop structures with solutions to $Q(GP_2)$. Such loop structures are captured by solutions of the looped query $Q(GP_1) \wedge Q(GP_2)$.

6.4.2 Query Generation Framework

The framework is depicted in Fig. 6.13. The first step identifies the graph patterns in datasets and extracts their corresponding instances. The input of this step includes the datasets and optionally some related constraints, such as size of the graph patterns. The output of this step is a set of pairs $\langle GP, I_{\mathcal{D}}(GP)\rangle$, where GP is a graph pattern and $\mathcal{D}$ is a dataset. The main challenge is that there can be *too many graph patterns with useless 0 correspondence*. To avoid generating such meaningless query pairs, we make sure that the shared variables of two queries (or graph patterns) belong to the same type. Hence, we perform data summarisation based on the types of nodes in the graph. Given an RDF graph, the summarisation is to generate a condensed description which reduces the search space of the graph pattern mining. Roughly speaking, the summarisation is an analogue to the schema, e.g. E-R diagram, in a relational database system. This summarisation takes the form of graph patterns which reveal the possible relations among individuals. One of its good properties is that any query which is interesting by our definitions is a subgraph (subgraphs) of the summarisation. This property allows jumping out from the "swamp" of original data graphs and focuses on the summarisations to mine interesting queries. Furthermore, the size of summarisation graph is extremely small when comparing it to that of

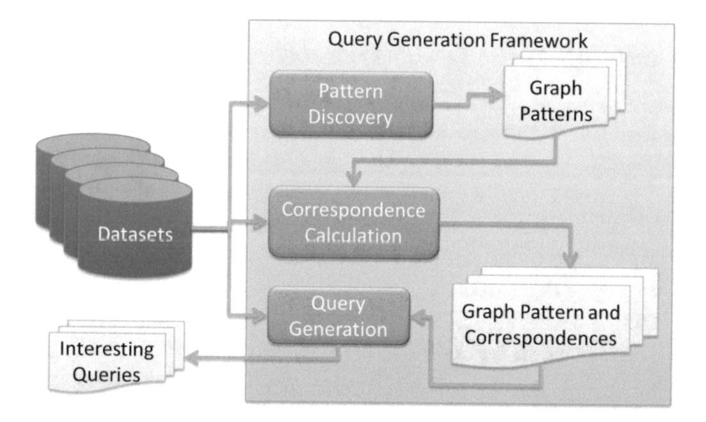

Fig. 6.13 Query generation framework

the original graph. The biggest summarisation graph in our test datasets only has 44 triples. Such tiny-sized summarisations can largely facilitate the mining process, e.g. expensive mining algorithms are applicable.

The second step computes the correspondences between graph patterns. The input of this step is the data summaries delivered by the previous step and the datasets. The output is a set of 4-tuples $\langle GP_1, GP_2, \mathit{support}, \mathit{confidence}\rangle$ where GP_1 and GP_2 are graph patterns, *support* and *confidence* are used to characterise the correspondences between GP_1 and GP_2. Assuming GP_1 and GP_2 share a vector of variables υ, then $\mathit{support} = \mid C_\upsilon(GP_1, GP_2) \mid$, $\mathit{confidence} = \frac{|C_\upsilon(GP_1,GP_2)|}{|I_\upsilon(GP_1)|}$ when $I_\upsilon(GP_1) > 0$ and 1 when $I_\upsilon(GP_1) = 0$. *support* indicates how frequent do GP_1 and GP_2 share instances w.r.t. υ. The higher *support* is, the more frequent the two graph patterns share instances on υ in general. *confidence* indicates how frequent do instances of GP_1 w.r.t. υ are also instances of GP_2. The higher $\mathit{confidence}(GP_1, GP_2)$ is, the more frequent that instances of GP_1 can be shared with GP_2. The challenge in this step is to *identify different patterns with desired correspondence*. In the domain of ontology learning, algorithms of inductive logic programming (ILP), association rule mining have been explored to find relationships between concepts and relations [141, 247]. Inspired by these works, we examine three different approaches (worst-case polynomial time) to find the corresponding graph patterns. **FOIL** (First-Order Inductive Learning) constructs the graph pattern by including a set of possible reachable variables via gradually growing the graph pattern. The algorithm selects the best variable from the set by a gain function (cf. [196]). FOIL tends to generate star-shaped graph patterns. **COMB** and **LOOP** approaches utilise the association rule mining technology. They tend to generate chain-shaped or looped graph patterns.

This third step generates insightful queries based on our discussion in Sect. 6.4.1. The input of this step is the datasets, and the computed correspondences between graph patterns. The output of this step will be a set of insightful queries or query pairs.

6.4.3 Evaluation of the Query Generation Method

We implemented the framework in Sect. 6.4.2 and evaluated its performance with benchmark datasets: **LUBM** (Lehigh University Benchmark) is an artificial dataset in which data is automatically generated. Its transparency makes it easier for us to examine whether the queries generated by our system are useful or not. We generated 15,247 triples in our evaluation. **DBLP** is a large and real-world dataset which includes bibliography data. In our evaluation, we used DBLP2011 data[18] (3,584,734 triples). **DBTune** hosts a selection of music-related RDF datasets. In our evaluation, we used the Jamedon[19] (1,047,950 triples) and BBC-PEEL[20] (271,369 triples)

[18] http://law.di.unimi.it/webdata/dblp-2011/.

[19] http://dbtune.org/jamendo/.

[20] http://dbtune.org/bbc/peel/.

Table 6.1 Query examples

	Query correspondences

1	`SELECT ?x`
	`WHERE {?x rdf:type lubm:ResearchGroup.}`
	
	`SELECT ?x`
	`WHERE {?x lubm:subOrganizationOf ?y.`
	`       ?y rdf:type lubm:Department.}`

	Query exceptions

2	`SELECT ?x`
	`WHERE {?x lubm:headOf ?y.`
	`       ?y rdf:type lubm:Department.`
	`FILTER NOT EXISTS`
	`       {?x rdf:type lubm:FullProfessor}}`

	Looped queries

3	`SELECT ?x ?y ?z ?o`
	`WHERE {?x lubm:worksFor ?z.`
	`       ?x lubm:teacherOf ?y.`
	`       ?o lubm:memberOf ?z.`
	`       ?o lubm:takesCourse ?y.`
	`       ?o lubm:advisor ?x.}`

datasets. They are both using the music ontology[21] and the FOAF ontology[22] as terminologies.

We applied our framework on the above datasets and generated queries on graph patterns with the strongest support and confidence. Such generation can be performed efficiently. We examined the top 20 generated queries (query pairs) for each dataset and the results showed that they were all meaningful. Some examples of generated queries from the LUBM dataset are presented in Table 6.1.

For example, query 1 suggests a high correspondence between the suborganisations of some department and research groups. This reveals that a suborganisation of a department is very likely to be a research group. Such insights will be helpful when investigating the administration structures in universities. Query 2 investigates if the head of a department must be a full professor. In the evaluated dataset, this indeed is the case. By automatically generating a query on cases where the head of the department is not known to be a full professor, users can easily identify potential exceptions. Another category of insightful queries are queries with loops, such as Query 3 in Table 6.1. This query suggests very high support and confidence that an advisor and an advisee work for (is a member of) the same entity, and the advisor is the teacher of some course that is taken by the advisee. This query contains three loops and is very difficult to be expressed with normal ontology language.

[21] http://musicontology.com/.

[22] http://www.foaf-project.org/.

6.4.4 Conclusion and Future Work

This section presented a novel and tractable approach to generate candidate insightful queries for knowledge graphs. A combination of data summarisation and different mining technologies has been exploited to extract graph patterns and construct candidate insightful queries. The evaluation shows that the proposed framework can generate insightful queries from synthetic and real-world datasets.

In addition to QG based on systematic parameterisation [92], it might be worth exploring how to extend the described framework in order to embed the support of tractable reasoning and schema [186], so that, on the one hand, reasoning can be done on the fly with data summarisation and on the other hand, reasoning can be used to eliminate query pairs that are inferred to share all (or no) solutions.

Chapter 7
Question Answering and Knowledge Graphs

Alessandro Moschitti, Kateryna Tymoshenko, Panos Alexopoulos, Andrew Walker, Massimo Nicosia, Guido Vetere, Alessandro Faraotti, Marco Monti, Jeff Z. Pan, Honghan Wu and Yuting Zhao

In the Digital and Information Age, companies and government agencies are highly digitalised, as the information exchanges happening in their processes. They store information both as natural language text and structured data, e.g. relational databases or knowledge graphs. In this scenario, methods for organising, finding and selecting relevant information, beyond the capabilities of classic Information Retrieval, are always active topics of research and development. *Question Answering*, i.e. retrieving exact and concise answers to questions asked by user in natural language, is one of the most promising among such methods.

Knowledge graphs play a key role in question answering. On the one hand, they are the natural encoding for structured knowledge extracted from texts, databases or other sources, making it available for efficient queries. On the other hand, they provide support to processing textual data. As such, they seem to be indispensable

A. Moschitti (✉) · M. Nicosia
University of Trento, Via Sommarive, 9 I-38123 POVO, Trento, Italy
e-mail: moschitti@disi.unitn.it

M. Nicosia
e-mail: m.nicosia@disi.unitn.it

K. Tymoshenko
Trento RISE, Povo di Trento, Via Sommarive 18, Trento, Italy
e-mail: K.tymoshenko@trentorise.eu

P. Alexopoulos
Expert System, Prof. Waksman 10, 28036 Madrid, Spain
e-mail: palexopoulos@expertsystem.com

A. Walker · J.Z. Pan
University of Aberdeen, King's College, Aberdeen AB24 3UE, UK
e-mail: andrew.walker.05@aberdeen.ac.uk

J.Z. Pan
e-mail: jeff.z.pan@abdn.ac.uk

J.Z. Pan et al. (eds.), *Exploiting Linked Data and Knowledge Graphs in Large Organizations*, DOI 10.1007/978-3-319-45654-6_7

in many application scenarios. In this chapter, we first introduce the tasks of question answering over text documents (Sect. 7.1) and knowledge graphs (Sect. 7.2), for which we present an overview of the relevant methodologies, technologies and systems. In Sect. 7.3, we further explain how knowledge graphs are used in the well-known IBM Waston DeepQA pipeline. Moreover, in Sect. 7.4, we describe a state-of-the art question answering system that combines knowledge coming from the text analysis and knowledge graphs.

7.1 Question Answering over Text Documents

This section provides a brief introduction to Question Answering (QA) systems over natural language text documents.

It gives a general discussion about the technical approaches that are usually used in implementing a QA system. Two of these approaches, are further elaborated in this part, which are Natural Language Processing (NLP) and Answer Ranking, respectively, are further elaborated in this part.

7.1.1 Realising a QA System: Approaches and Key Steps

Question answering (QA) systems differ from traditional search engines because they accept in input a question in the natural language and output the single information answering the question, or in substitution the passages or paragraphs, which may contain the answer to the question. Providing a direct answer or passage aims to reduce the time necessary to find and validate the relevant information.

If a question in the natural language is used as query for a search engine it may happen that the answer will be contained in documents from the first result page, or even in text snippets associated with search results. This is mainly due to the terms'

G. Vetere · A. Faraotti
IBM Italia, via Sciangai 53, 00144 Rome, Italy
e-mail: gvetere@it.ibm.com

A. Faraotti
e-mail: alessandro.faraotti@it.ibm.com

M. Monti · Y. Zhao
IBM Italia, Circonvallazione Idroscalo, 20090 Milan, Italy
e-mail: marco.monti@it.ibm.com

Y. Zhao
e-mail: yuting.zhao@it.ibm.com

H. Wu
King's College London, De Crespigny Park, London SE5 8AF, UK
e-mail: honghan.wu@kcl.ac.uk

similarity between the question and the text containing the answer. While IR systems can be used as baseline QA systems, the technology employed by such systems is not tuned to question answering. These systems ignore the syntactic relations between questions and answers, and do not perform any analysis of the natural language in the documents.

The QA systems have applications in different scenarios. They can be used to efficiently find information on the Web or in specific document collections. Since they can be tailored to a given domain there is interest in their usage in medical and legal fields or in specialised knowledge bases owned by companies, agencies, public administrations and research centres. Moreover, they are very effective in finding information when every document in the collection is equally important and the hyperlinks between them can be neglected. Modern QA systems employ methods and theories coming from different fields: Information Retrieval, Natural Language Processing and Machine Learning.

The retrieval step is at the base of every QA system. Its goal is to retrieve the documents related to the question which contains the answer. Maximising the number of retrieved documents maximises the probability of finding the answer and thus the recall, but considering too many documents introduces irrelevant information which must be discarded. The document collection used as source is very important because if the answer to a question is not contained in it, the QA system has no chance of returning a meaningful answer.

Machine learning (ML) is a branch of artificial intelligence focused on algorithms capable of learning relationships in the training data and use this experience to generalise and perform a specific task on unseen data. The performance of the algorithm at the given task improves with the increase of experience. Thus, more training data leads to better learned models and improved performance. ML is relevant to QA systems because it is used in most Natural Language Processing tasks and re-ranking.

Natural Language Processing (NLP) is a multidisciplinary research area in the field of computer science, Machine Learning and linguistics. It concerns the process of automatically parsing text (syntactically and/or semantically) with the aim of extracting, analysing and understanding the information it contains and generating new information also in the text format. The analysis performed on the text has different stages of complexity, e.g. basic processes are: tokenisation, lemmatisation, morphological, syntactic and semantic analysis. Early work in NLP widely used rule-based methods, whereas nowadays Machine-Learning algorithms are applied for training linguistic models from data. The data typically contains examples of correct versus incorrect output of the NLP function that the Machine-Learning algorithms are supposed to replicate on unseen data. The NLP includes different tasks; some of them are Tokenisation, Sentence Boundary Disambiguation, Named Entity Recognition, Part-of-speech Tagging, Chunking, Parsing, Relation Extraction, Semantic Role Labelling and Co-reference Resolution. These tasks are carried out as preliminary steps in the design of applications dealing with a natural language, including QA systems.

The rest of this subsection will give a brief introduction of the two main steps in realising a QA system, i.e. the NLP and the Ranking.

The NLP Step

As mentioned earlier, the NLP step is usually composed of a list of tasks which are often related to one another. A task can rely on the output of the analysis performed by another task, or in some cases two tasks can benefit one of the results of the other, and thus they can be re-executed in a sort of loop. When it is possible to avoid duplicate work and analysis, the different NLP tasks can be arranged in a sequence where the latest tasks use the results already made available by previous tasks. Usually, basic tasks such as tokenisation and sentence boundary detection, which are frequently used by more complex tasks, are carried out once first, and their results are readily shared with subsequent jobs.

NLP Pipelines are used to wrap, compose and orchestrate the different tasks and their inputs and outputs, and facilitate their reuse. A pipeline is a sequence of components performing different NLP tasks. Pipelines perform specific sets of tasks, their internal flow is conditional, and thus based on the analysed data and the results of the analysis. Moreover, they can be reused in a different application. The components can be plugged or unplugged easily, making the general system extensible and flexible. The main advantage of an NLP pipeline is that it separates the analysis logic from the parts of the application exploiting analysis results. In this way, pipelines of arbitrary complexity can be instantiated for different applications, and shared.

UIMA (Unstructured Information Management Applications) is a framework for building systems composed of components whose interfaces are defined in terms of input and output. Every component carries out a specific task and produces results, which are collected in a unique data structure and can be used by other components. The configuration aspects are mainly managed by XML files. The framework manages the components and the data flow between them, it is scalable and parallelisable and it is appositely developed to be used for building applications which analyse a huge quantity of unstructured text. UIMA is used in the DeepQA project by IBM and powers *IBM Watson*, the QA system that in 2011 beat the human champions of *Jeopardy!*, an American quiz television show.

The typical tasks in an NLP pipeline or framework are summarised as follows.

- **Tokenisation** is the process of converting text into meaningful elements called tokens. Tokens can be words, phrases or symbols. Usually, they are produced by separating the words in the text using whitespaces. However, this approach is not sufficient for complex languages or when tokens should be formed by multiple words. Thus, the tokenisation process often includes additional heuristics, language-specific models and lookup tables. The output of tokenisation is used by the other NLP tasks.
- **Stemming** is the process of reducing words into stems. A stem is the part of a word common to all its inflected variants. Affixes can be attached to the stem in order to produce all its inflected forms. Stemming does not use the context of words in the text, operates individually on single words and uses heuristics to remove affixes. Not every stemmed word is still a valid word. Stemming is prone to ambiguities because many words having unrelated meanings may share the same stem.

- **Lemmatisation** is the process of matching words with their lemmas. A lemma is the canonical form or the dictionary form of a set of words. Lemmatisation exploits the context around words in the text and uses the part-of-speech information to discriminate between words with different meanings. For these reasons it is slower than stemming and produces different results.
- **Sentence Boundary Disambiguation** is the process of determining the beginning and end of sentences in text. Delimited sentences are needed for other NLP tasks. The punctuation marks are ambiguous identifiers of sentence endings: periods can occur in abbreviations, numbers and ellipsis, while question and exclamation marks can occur in quotations and other artefacts of written language. Thus, to obtain the best performance, it is not sufficient to consider only punctuation, and approaches including heuristics and rules learned from training data are used.
- **Named Entity Recognition** (NER) or entity extraction is a subtask of information extraction. A named entity recogniser extracts entities of a given type from an unstructured text. Such entities can include persons, organisations, locations, abbreviations, numbers and dates, but are not limited to these categories. The NER task is domain dependent: every domain has its own entity types. Recognisers can be trained on specific domains to discover specific kind of entities. In general, they perform well in their focus domain, but are not so effective in others. Examples of domain-specific NERs can be seen in the biomedical field, where they can recognise DNA, RNA and proteins.
- **Part-of-Speech Tagging** or POS Tagging is the process of assigning parts of speech to words in a text, considering the characteristics and roles of the different parts of speech and the context around words. Relationships between adjacent words aid the tagging. Some examples of parts of speech are nouns, verbs, articles, adjectives. Automatic POS Tagging employs from 50 to 100 parts of speech, which can also be referred to as word classes or lexical categories.
- **Chunking** or shallow parsing adds to POS tags additional information on the constituents of a sentence. Constituents can be noun groups, verbs or verb groups. Chunking produces a partial syntactic structure of the sentence and is more efficient and robust than full parsing. The structure is represented as a tree, which usually is two levels deep. The leaves of the tree contain the words and their POS tags. The upper level groups the words into constituents. The use of shallow parsing is preferred to parsing when the full parse tree is not needed.
- **Syntactic Parsing** produces the syntactic parse tree of a sentence. The tree contains information about the grammatical structure of the sequence of words in the sentence, and the syntactic relations between them. Often parsing is ambiguous and for a single sentence there are multiple representations. In these cases only the most probable parse tree is kept. Parse trees are important features for different tasks when working on language, among which is question answering.
- **Relation Extraction** (RE) is a subtask of information extraction and it is useful to discover connections between entities, to create new knowledge bases or to augment existing ones. The RE supports also question answering when the question asks for something related to a given entity and it can be mapped into a query which can be issued to a knowledge base.

- **Semantic Role Labelling** (SRL), also called shallow semantic parsing, is the process of identifying the semantic roles in a sentence. Semantic roles define the functions of participants to events contained in a sentence. Identifying the semantic roles is useful because language allows us to express the same fact or event in different ways, even if the participants are the same and so are their roles. The typical case is expressing a fact using an active or a passive voice: even if the syntax is different, the meaning of the sentence does not change. The SRL has applications in Information Extraction, Question Answering, Summarisation and Machine Translation.
- **Co-reference Resolution** is one of the important steps of Discourse Parsing. It is the process of finding the parts of discourse in a text referring to the same entities or things, which is very important for gaining a better understanding of the text. An entity or thing can be referenced using different forms: synonyms, pronouns, multiple names or figures of speech. Moreover, when the current sentence contains a reference, the reference best representing that individual may be in another sentence, and thus the reader, and subsequently the resolution tool, must go back or forward in the text to find it. This issue makes the co-reference resolution task challenging.

The Ranking Step

Ranking is the main problem in many Information Retrieval and Natural Language Processing tasks, such as document retrieval and question answering. The purpose of ranking is to order a list of objects using their features with the goal of putting the more relevant objects first in the list. The relevancy of an object can be subjective and depends on the problem at hand and the available data.

In document retrieval which includes Web search the user typically enters a query, the documents in the system are ranked and only the top ranked are presented. In this case, the ranking considers the relevancy of the documents with respect to the query.

For document retrieval a more formal definition of the ranking problem is the following: given the document collection D and the query space Q, the ranking function $rank : D \times Q \Rightarrow \mathbb{R}$.

In question answering the query is in the form of a question. A retrieval step returns the passages which may contain the answer, and the passages or the generated candidate answers are ranked exploiting NLP techniques, Machine Learning and supporting evidence.

Learning to Rank

The performance of ranking models usually depends on several parameters. They can be manually tuned, but this process can be difficult when they are in high number and can cause overfitting. In the literature, different models are proposed, each of them with its own strengths and weaknesses. Issues rise when combining them in order to get a model which is more effective than the single models because the process is non-trivial. Machine Learning is useful in these cases to automatically learn the model parameters, to combine the results coming from different models and to avoid overfitting.

The methods using Machine Learning to perform ranking are called **Learning to rank** (LTR) methods. The general process performed by learning to rank methods concerns collecting training data, such as queries and associated labelled documents, performing feature extraction on them, learning a ranking model by minimising a loss function on the training data and using the learned model to infer the ranking of new data.

The ranking model in learning to rank methods is feature based. Features are extracted from the elements to rank, such as a query or question and the associated documents, and put in a vector. This approach permits combining the results of different retrieval models by including their outputs as features. A query-document TF-IDF score and BM25 score, along the document length, PageRank [40], HITS authority, and hub values [134] can be candidate features to put in the vector. Features should be significant and ideally enclose the characteristics which affect the ranking. Features allow for generalisation, indeed the model is trained on a limited set of data but it is also able to perform on novel query and documents.

Most of the state-of-the-art algorithms learn the best way to combine features extracted from query and document pairs using discriminative training [147]. In this kind of training, the correctness of a model is improved through an objective function that penalises parameter values leading to errors. Thus, as new training examples are read, the model parameters are tuned in order to fit them and reduce misclassification.

Discriminative training is based on four pillars:

1. the input space, which contains the training and test data represented as feature vectors;
2. the output space, which can be of the task—for example -1 or $+1$ in the case of binary classification—or can facilitate learning—for example a confidence value about the class membership of an example;
3. the hypothesis space, containing the functions mapping from the input space onto the output space;
4. the loss function, which is a function mapping a prediction generated by the hypothesis onto a real number representing the cost associated to that prediction. It is a penalty for incorrectly classifying examples.

The loss function is an important characteristic of learning to rank algorithms which can be divided into pointwise, pairwise and listwise approach algorithms.

Pointwise algorithms do not see the list to rank as a whole but break it into its components, namely the single documents. Each query and document pair can have a relevancy label, or an ordinal or numerical score. The classifier must predict this score for unseen pairs. Thus, the ranking problem is approximated by a regression problem. The loss function computes the difference between the model output for an example and a prototype notion of pure relevancy learned from training data.

Pairwise algorithms consider pairs of documents in the list to rank, reducing ranking to pairwise classification. Given a pair of documents, the classifier decides which one is more relevant to a query.

Listwise algorithms consider the list to rank in its entirety, without breaking it into single documents or into document pairs. The algorithm tries to minimise a loss function which compares the predicted permutation of documents with the true ranked list [255].

Evaluation of Ranking

A ranked list can be evaluated having a set of documents associated to the query that produced the list, and a list of relevance judgements for each document with respect to the query. Then, the ranked list can be evaluated using some measure. The ranking system's performance should be assessed not on a single query but on a set of queries, averaging the values of the evaluation measure.

In the TREC(Text Retrieval Conference)[1] competitions, the participating systems select from a corpus the possible relevant documents for a given query. Then, the top 100 documents retrieved by all the systems are judged by human assessors using a binary or ordinal scale, and pairwise preferences, e.g. document A is more relevant than document B. Eventually, the documents are arranged in a totally ordered list for each query.

A popular evaluation measure is the Precision at Position k which tells how many relevant documents are in the first k position of the list retrieved for a given query, and it is usually computed for different values of k.

$$P \bullet k = \frac{number\ of\ relevant\ documents\ in\ top - k\ results}{k}$$

Another measure is the Average Precision:

$$AP = \frac{\sum_k P \bullet k \times l_k}{number\ of\ relevant\ documents}$$

where l_k is equal to 1 if the document at rank k is relevant and to 0 otherwise. The mean average precision (MAP) is the AP averaged over all queries.

A measure used for evaluating QA systems is the mean reciprocal rank (MRR), which is averaged on all the questions and is higher if the first relevant answer is at the top in the retrieved lists.

$$MRR = \frac{1}{|Q|} \sum_{i=1}^{|Q|} \frac{1}{rank_i}$$

[1] http://trec.nist.gov/.

7.2 Question Answering over Knowledge Graphs

In this section, we consider question answering (QA) systems where the data is stored as a knowledge graph, according to some ontological schema and with each node connected to others by various relations. These systems (also called ontology-based QA systems) operate on highly structured data where the semantic details are explicitly defined within the context of the ontology and typically also with references to external ontologies. Natural language queries however are structured only insofar as one word comes before another, and the semantic content is often unclear. Thus, the challenge is the accurate and efficient interpretation of these queries as they pertain to the knowledge contained in the relevant ontologies.

More specifically, the main task to be tackled is the transformation of the natural language query into a set of required RDF[2] triples, typically expressed in the SPARQL[3] query language and in accordance with the system's ontology. Achieving such a transformation, answering of the query is simply performed by executing it against the ontology and showing the results to the user. Thus, for example, the question *"Give me the birthdays of all actors of the television show Charmed"* against the DBPedia[4] ontology would be transformed into the following SPARQL query:

```
PREFIX dbo: <http://dbpedia.org/ontology/> PREFIX
res:<http://dbpedia.org/resource/> SELECT DISTINCT ?date
    WHERE {
    res:Charmed dbo:starring ?actor .
    ?actor dbo:birthDate ?date .
}
```

Towards performing this task, several question answering systems for semantic data have been proposed by the scientific community, including Aqualog [149], PowerAqua [148], NLP-Reduce [6], Pythia [236] and FREyA [63]. These systems vary in several aspects like, for example, domain specificity (some are domain-specific whereas others are schema-agnostic) or methods employed (e.g. deep linguistic analysis versus statistical approaches).

In what follows we provide a brief overview of the existing ontology-based QA tools and frameworks and some architectural and methodological guidelines as well as the best practices about the effective implementation of such systems in enterprise scenarios. As we will explain below, enterprise QA scenarios may have different requirements and/or characteristics than, for example, open-ended scenarios regarding the whole Web.

[2]http://www.w3.org/RDF/.
[3]http://www.w3.org/TR/sparql11-query/.
[4]http://dbpedia.org.

7.2.1 State-of-the-Art Approaches for Question Answering Over Knowledge Graphs

Ontology-based question answering systems can be categorised along two main dimensions: (i) the degree of domain customisation they require in order to be effective and (ii) the subset of natural languages (NL) they are able to understand (full grammar-based NL, controlled or guided NL, pattern based). At one end of the spectrum, systems are tailored to one or more a priori defined domains and most of the customisation has to be performed or supervised by domain experts [54, 76, 236]. At the other end, the customisation is done on the fly through user interaction and adaptive learning [63, 149]. A third categorisation dimension is the ability of the system to scale to the open Web (i.e. Linked Data) without being restricted to a limited set of domains. Relevant approaches in that category include FREyA [63] and PowerAqua [148].

In particular, the goal of FREyA is to develop user-friendly interfaces to Linked Open data. They emphasise the use of clarification dialogues with the users when it encounters as-then unresolvable ambiguities from which it trains itself further for future queries. The general process they follow consists of the following steps:

1. **Identification of Ontology Concepts (OCs)**: Terms in the NL query are mapped to URIs from the ontology. In cases where multiple URIs could be referring to (e.g. Mississippi to the state or the river) it will try to disambiguate from the context (e.g. "which rivers flow through Mississippi?" is unambiguously referring to the state). If this fails it will engage in dialogue with the user to generate clarification.
2. **Identification of Potential Ontology Concepts (POCs)**: Terms in the NL query tagged as noun phrases, or adjectives within a noun-phrase, are labelled as POCs. These POCs are then compared with the OCs for syntactic overlap, such that if POC is automatically labelled as the OC. If no such overlap exists the user will be engaged to select the appropriate OC from a list of suggestions.
3. **Generation of OC Suggestions**: The closest OC is found by walking the syntax tree and ontology reasoning is used to generate suggestions for the POC. "None" is always included as a suggestion to enable the user to ignore them. If the OC is a class or instance, suggestions will be class linked to it by 1 (and only 1) property, along with all of the properties for the OC. If the OC is a numeric property, the suggestion will be aggregate functions such as maximum, minimum and sum. Finally, if the OC is an object property it will suggest the domain and range classes of the OC.
4. **Ranking Suggestions**: Suggestions are ranked by a variety of mechanisms including string similarity with Monge–Elkan metrics and the SoundEx algorithm, and synonymy as determined with WordNet and Cyc.
5. **Generation of SPARQL Query**: Once the query is resolved into a set of OCs, the "joker" elements are inserted to form complete triples (or triple chains) and the triples converted to SPARQL that returns all of the relevant entities. Identifying and presenting the required information is done later.

6. **Answer Type Identification**: The algorithm for identifying the answer type is an independent paper. It involves an algorithm over a syntax tree generated by the Stanford Parser to extract either the focus of the question (a noun-phrase) or what they call an "answer type identifier" (ATI; e.g. a *wh*-question). By identifying the first OC (FOC) and consolidating it with this extraction they identify a number of possible answer types. If there is just one, this is identified as the answer type; otherwise the suggestions (plus "none") are presented to the user.

FREyA relies on the interaction with the querent for most queries, but this is an inherent cost of a system intended to give its users full control over the querying process. Suggestions for querent-driven disambiguation are complex to derive and the process will scale poorly. The generated SPARQL returns all variables, and extraction of the answer occurs afterwards, possibly incurring a bandwidth cost for remote data that could be minimised by formulating queries in the first place only to return the relevant information.

On the other hand, the focus of PowerAqua is the integration of information from multiple, heterogeneous semantic resources to derive answers to natural language queries. This requires, first, efficient and accurate identification of which semantic resources are even useful for answering the query. Further, due to the inevitably extensive volume of data potentially being handled, they consider scalability issues, and the issues inherent in third-party sources of noisy and incomplete data. The query answering process applied here is the following:

1. **Query-triple extraction**: A query undergoes linguistic processing to be converted to a set of triples, $< subject, predicate, object >$ allowing for ambiguous terms (e.g. $< subject1/subject2, predicate, object >$) and unknowns (e.g. $< subject, ?, object >$).
2. **Element mapping**: Initially PowerAqua simply collects all potentially relevant semantic entities by lexical relations to the query terms and their synonyms and hypernyms (from WordNet). The Semantic Web itself is also used to collect any others via the `owl:sameAs relation`. As this will clearly often yield multiple candidates per term, a subsequent step is required to disambiguate between possibilities.
3. **Triple Similarity Service**: In order to convert the mapped triples to sets of semantic triples from which to derive answers, PowerAqua uses simple strategies initially and incrementally extends them if they prove insufficient. It initially selects ontologies that cover the most of the user query, only examining others when/if a solution cannot be derived. It favours mappings covering an entire compound (e.g. "Clint Eastwood" to $< http : //mpii.de/yago/resource/Clint_E astwood >$) over mappings covering subsets of the query triple. Finally, it favours shorter relations between candidate entities (ideally direct relationships) over those requiring mediating concepts.
4. **Response ranking**: The results are ranked according to three distinct confidence measures: (a) the mapping on ontological facts, (b) the disambiguation algorithm on the interpretation of the answer and (c) how the information from the various ontologies is merged together.

PowerAqua has been evaluated with 69 questions compiled by 7 users familiar with the Semantic Web such that the answer could certainly be found in at least one of 700 semantic documents (3 GBs of metadata). Accuracy (percentage of questions answered correctly by the system) came to 69.5 %. They further analysed the failures to determine where the errors occurred, finding 7.2 % originated with the linguistic analysis of the query, 18.8 % from either not finding or discarding valid element mappings and 4.3 % from incorrectly locating the ontology triples to answer the query. The time required by the system to answer queries ranged from 0.5 to 79.2 seconds, averaging at 15.39 s.

From the approaches not targeting the open Web, Pythia [236] takes a grammar-centred approach by building up an ontology-dependent grammar with the help of which it parses the natural language queries of users. The ontology is enriched manually with information about its verbalisation (using LexInfo) and from this an ontology-specific grammar is automatically generated (to augment an ontology-independent grammar). This grammar is used to initially parse the NL query, which avoids the problems inherent in post-parse analysis, conversion and tagging. The grammar should handle morphological variations of ontological entries and passive/active voice, but this depends on the quality of the lexical enrichment.

From the Mooney geoquery set of 880 questions they annotated 865 that were covered by a subset of DBpedia with FLogic query results. Approximately 2.5 h were spent compiling the LexInfo model on the ontology, achieving precision of 82 % and recall of 67 % (F-measure of 73.7 %). They observe that some errors were due to questions being either ill formed (either syntactically or semantically).

The requirement for manual enrichment of the ontology with linguistic information, requiring 2–5 min per class/instance/relation, is an unwelcome bottleneck; though arguably an acceptable one, given its re-usability. Their parser requires queries to be grammatically correct (as per the provided grammars) in order to process them at all. The use of a more flexible parser might reasonably extend the scope of possible queries significantly.

On the other hand, AquaLog [149], a precursor of PowerAqua, does not rely on *a priori* manual linguistic enrichment of the ontology but uses (i) a linguistic component based on GATE (General Architecture for Text Engineering)[5] to extract, through manually defined pattern extraction rules, triples from the query text and (ii) an interactive relation similarity service to disambiguate between the alternative representations of the user query, based on the ontology taxonomy and relationships. When the ambiguity cannot be resolved by domain knowledge the user is asked to choose between the alternative readings. Furthermore, a learning component is used to automatically obtain domain-dependent knowledge by creating a lexicon, which ensures that the performance of the system improves over time, in response to the particular community jargon (vocabulary) used by end users.

[5]http://www.gate.ac.uk.

Another relevant system is QAKiS [43], which is primarily concerned with handling the multiple ways in which a relation can be expressed in a natural language query. By drawing from Wikipedia infoboxes and texts they attempt to derive relation patterns with which to interpret queries. They tackle this issue through automatic capture and matching of relational patterns. As they focus on DBpedia, compiled via the infoboxes on Wikipedia articles, and assume that wiki content will include textual representation(s) of the infobox content, relational patterns are extracted by finding sentences involving both the subject and object of the relation and examining their syntactic structures.

QAKiS's limitation is that its technology relies heavily on Wikipedia, from which DBpedia is derived. In order to use the technology with another knowledge source a similar corpus of text containing the ontological information must be provided. Furthermore, QAKiS only handles queries of a specific form: those expressing only one relation between the subject (which must be a Named Entity according to the Stanford Parser) and the object.

7.2.2 Question Answering in the Enterprise

In enterprise information access scenarios, QA systems are aimed to replace traditional keyword-based search interfaces and allow users to exploit more effectively the structured knowledge in whose creation the enterprise has invested. This is probably the case for non-enterprise scenarios, such as Web search, as well. There are however some important differences between enterprise knowledge graphs and non-enterprise ones as listed below:

- **Enterprise knowledge graphs are (expected to be) of higher quality than non-enterprise ones**. This is because enterprise graphs are created and maintained in a controlled, centralised way with particular focus on quality aspects such as consistency, richness of ontology schema, timeliness of data, etc. On the other hand, many Linked Open Datasets that are available on the Web do not follow equally disciplined processes, the result being, for example, schemaless datasets that are hard to use within a QA system. This makes some challenges related to the development of QA systems (e.g. having to guess the data schema) less relevant/important for the enterprise case.
- **Enterprise knowledge graphs are less diverse and less heterogeneous**. Homogeneity is ensured by the centralised development process mentioned above while low diversity is a natural consequence of the fact that an enterprise is primarily interested in modelling knowledge about its business activities and environment rather than the whole world. Obviously this domain specificity can help build more effective QA systems.
- **Enterprise users can be more predictable than Web users with respect to the structure and content of the questions they ask**. That is because enterprise users will use the QA system for a particular task/goal related to their job and thus the

questions they will want to be answered will be less open-ended and more focused on the enterprise's domain. This is another factor that can make enterprise QA systems more effective than their non-enterprise counterparts.

- **In enterprise scenarios, answering some types of questions reliably and consistently is more important than answering all types of questions but inconsistently**. While research QA prototypes should strive to support an as large as possible range of question types even if effectiveness is low, in enterprise settings the reliability and predictability of the system's performance comes first. This means that enterprise QA system developers should follow a bottom-up approach by tackling first "simple" question patterns that provenly work well and only then move towards more complex patterns.
- **Transparency and explicability of the QA process is important**. As QA systems are still very far from being 100 % accurate, it is important that whenever a question cannot be answered, a quick and clear explanation of why this happened should be made available to the administrators/developers of the system and of the knowledge graph.
- **It should be possible for the QA process to be improved by means of user feedback**. Enterprise users can be patient with a non-fully accurate system as long as this looks able to improve over time. Therefore, enterprise QA systems should implement feedback-based or other types of effectiveness improvement mechanisms.

Now, the typical pipeline that QA systems implement consists of four stages, based on NLP tasks, Sect. 7.1.1:

- **Stage 1: Question Linguistic Analysis**. In this stage, the natural language question is analysed and transformed into an intermediate representation that reflects the potentially intended interpretation(s) of the question in a more structured way. This representation, as we will explain below, can range from a simple syntactic tree to a complex query template.
- **Stage 2: Entity Mapping and Disambiguation**. In this stage, the question's terms are mapped to entities (concepts, relations, attributes, instances) of the ontology. A typical problem to be resolved here is ambiguity as a term may map to more than one entity in the ontology. In any case, the result of this stage is typically an updated set of question interpretations with the main linguistic terms having been replaced by their ontological counterparts.
- **Stage 3: Formal Query Construction**. In this stage, the interpretations of stage 2 are transformed into formal queries that can be executed against the knowledge graph. For RDF graphs these queries would be expressed in the SPARQL language.
- **Stage 4: Query Execution and Answer Provision**. The final stage executes the query (or queries) produced by stage 3 against the knowledge graph and uses the retrieved data as an answer to the original user question. In this stage, it is also where the user may interact with the system and provide feedback.

From the above stages, the first two are the most complex and challenging ones, practically due to the ambiguity and informality of the question's natural language representation. For this reason, there are many different approaches and methods for performing these two tasks, each with different advantages and disadvantages.

A first set of approaches are those based on controlled natural language approaches like, e.g. GiNSENG [29]. These typically consider a well-defined restricted subset of natural languages that can be unambiguously interpreted by a given system. Nevertheless, it can be hard for non-expert users to learn such languages. Another set of approaches perform the linguistic analysis of the question in a domain-independent manner without considering the available domain ontology. For instance, the linguistic analysis component of the PowerAqua system transforms the question into a set of linguistic query triples, in the form of <subject, predicate, object>, by using manually crafted JAPE (Java Annotation Patterns Engine) grammars based on the GATE framework. In this way, for example, the question *"Which startups are the competitors of Orange?"* is transformed into the triple *<startups, competitors, Orange>*. The idea is that these triples are easier to be mapped to the domain ontology which follows the same triple-based representation.

A different transformation is supported by template-based approaches, such as LODQA [131] and TBSL [235], which construct a query template (or pseudo-query) on the basis of a linguistic analysis of the input question. For example, the template for the question *"Which startups are the competitors of Orange?"*, produced by TBSL, is the following:

```
QUERY: SELECT ?y { ?y0 ?p1 ?y. ?y rdf:type ?p0. }

SLOTS: y0: RESOURCE {orange} p0: CLASS {startups} p1:
    OBJECTPROPERTY
{competitors}
```

Similar to the triples approach, the templates are generated by means of predefined template patterns. In both cases, the entities identified within the query triples or templates need to be subsequently mapped to entities from the domain ontology. This is practically an entity resolution problem, as defined in Sect. 4.3, and as such it can be tackled by means of the corresponding frameworks mentioned there.

Finally, there are question interpretation approaches that perform stages 1 and 2 in a joint manner by utilising directly the domain knowledge graph. These include:

- Graph exploration approaches [214, 232], that interpret a natural language question by mapping elements of the question to entities from the knowledge base, and then proceeding from these pivot elements to navigate the graph, seeking to connect the entities to yield a connected query.
- Machine-Learning approaches [25, 206] that attempt to learn semantic parsers using the knowledge graph and question/answer pairs as training data.

Although there is no clear "winner" among the above approaches as to which performs best in all cases, our view is that in enterprise QA scenarios, stages 1 and 2 of the pipeline should be clearly separated, i.e. the linguistic analysis of the question should be performed in a domain-independent form. The rationale for this is that transparency and explicability of the QA process in an enterprise scenario is important in order to be easier for the administrators/developers of the system and/or the knowledge graph to pinpoint problems and improve the system. In that sense, it is important to know if the system failed to interpret a question correctly because the linguistic component failed to parse the question correctly or because the knowledge graph did not contain one or more entities mentioned in the question. In the former case, one would know that the linguistic component should be improved while in the latter case it would only be necessary to enrich/expand the knowledge graph. With joint approaches, this kind of error analysis would be more difficult.

7.3 Knowledge Graph and Watson DeepQA

7.3.1 What Is Watson DeepQA?

In February 2011, US television aired the *"IBM Challenge"* a man-versus-machine competition in which a featured computing system named *Watson* played and won a two-game match against Ken Jennings and Brad Rutter, the biggest all-time money winners in the *Jeopardy!* game history. The *Jeopardy!* game is one of the most famous American television shows about trivia on a wide variety of topics. Players receive tricky clues in the form of answers and must phrase their answers in question form. An example trivia is *"Involves money that does not have to be repaid, and generally awarded based on need."*, and the answer is *"What are grants?"*. Watson is a question answering (QA) system developed within the DeepQA project by an IBM research team led by David Ferrucci.

While phrasing an answer in question form is an easy task, finding the right answer is really difficult because of the complexity the clues can reach and the open domain they span. Some questions must be decomposed and solved in their different parts, others are puzzles and require specific steps for finding the answer. Watson employs several loosely coupled methods to generate candidate answers and uses their result to choose the answer with the highest confidence. The main aspects of Watson are briefly explained now, while a comprehensive overview of Watson and the challenges of building it are described in [78].

Open-domain QA is one of the most challenging problems in Artificial Intelligence and requires a combination of Information Retrieval, Natural Language Processing, Knowledge Representation, Reasoning and Machine Learning. All these technologies have been applied in developing the Watson pipeline shown in Fig. 7.1.

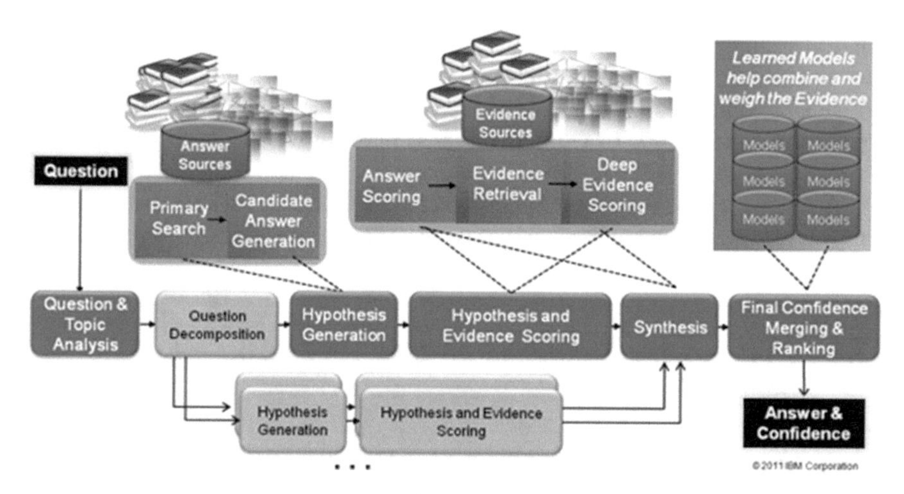

Fig. 7.1 Watson DeepQA architecture. *We are very grateful that the DeepQA Research Team (http://researcher.watson.ibm.com/researcher/view_group_subpage.php?id=2159) allows us to reuse their DeepQA Architecture diagram in this book

The Watson question-answering process is comprised of four main macrosteps: *Question Analysis, Hypothesis Generation, Hypothesis Evidence Scoring, Final Confidence Merging and Ranking.*

Jeopardy!'s trivia questions fall into categories which are useful as clues for finding answers, but when they are not very informative they can be misleading. Understanding what factual information the question is really asking is vital for the system so as to produce constraints for the answers. This task is named the Lexical Answer Type (LAT) determination. Usually, the LAT is a word in the clue indicating the type of answer but in the 12 % of clues it is not present and must be inferred from the context. Together with NLP shallow and deep parsing, the LAT determination task is included in the *Question Analysis* step aimed at understanding what the question is asking as to how the question will be processed. Results of the question analysis are exploited in the *Hypothesis Generation* step to produce candidate answers' snippets. This primary search is focused on recall and uses multiple search engines and data storages to retrieve as much the potentially answer-bearing content as possible. A variety of approaches are used to search including text search, document search and knowledge search. The search corpus is obtained expanding a base corpus: seed documents are selected and related documents are retrieved from the Internet; text nuggets are extracted from the document and only the most informative are added to the corpus. Additional kinds of structured and semi-structured content used are database, taxonomies and ontologies. Each candidate answer is considered a hypothesis whose correctness the system has to estimate with a degree of confidence. A soft filtering is also executed in this phase; lightweight scoring algorithms are used to filter the initial big set of candidate answers. A more resource-intensive filtering is performed in the *Hypothesis Evidence Scoring* step. Each remaining candidate is evaluated gathering additional supporting evidence and applying numerous deep scoring analyses. These results are collected and merged to choose the single best-supported hypothesis. Multiple surface forms of the answers are merged into a single form. The ranking of the hypotheses is performed using Machine-Learning techniques in the *Final Confidence Merging and Ranking* step. The goal of this phase is to identify the best-supported hypothesis among hundreds of candidate answers ranked with potentially hundreds of thousands of scores.

7.3.2 What Are the Knowledge Graphs Used in Watson DeepQA?

Knowledge represented in a structured format has been playing important roles in various question answering systems. A straightforward approach is to compile such structured knowledge by creating from scratch and/or extracting and curating from those in unstructured format. For example, in about a decade ago, Cyc system [59] produced a common-sense knowledge base, which was used for supporting passage retrieval and performing deductive QA. About five years later, when IBM was developing Watson DeepQA, the landscape of data science had changed significantly. Among others, the development of Semantic Web and Linked Data had brought in a healthy and growing ecosystem that facilitated the publishing, reusing and exploring structured data on the Web. This advancement brought new options for the DeepQA team.

In general, various types of structured data were used in Watson DeepQA. In this section, we mainly introduce the knowledge graphs reused and their extensions.

Off-the-shelf Knowledge Graphs

One of the key techniques used in Watson is the utilisation of Wikipedia knowledge for candidate answer generation [51]. Therefore, naturally, DBpedia was adopted, which is a knowledge graph version of Wikipedia.

In addition, due to the extensive spatial knowledge required in Jeopardy! challenge, Freebase knowledge base was also used as a complement to DBpedia. Instead of using the whole knowledge repository, a small portion of spatial information was extracted. In particular, the containment and border relations between geo-entities in Freebase provide higher precision and coverage than those in DBpedia.

The third off-the-shelf Knowledge Graph used in Watson DeepQA was the YAGO (Yet Another Great Ontology), which is a large taxonomy of more than 100,000 types with mappings to WordNet synsets. DBpedia instances are usually classified as low-level types in YAGO's class hierarchy. Travelling up to high-level types via YAGO's hierarchy will help Watson generate precise answer candidates. Furthermore, the relevant hierarchy structure is also very useful for matching answers that are using synonyms.

Extensions of Off-the-shelf Knowledge Graphs

To tackle Jeopardy! challenge, understanding the exclusivities between entity types is very helpful for ruling out incompatible answers. For example, if the question is asking about a *city*, the system can easily rule out those candidate answers whose types are *person*. However, YAGO does not provide the disjointness between types. Therefore, the first extension adopted by the Watson team was the generation of type disjointness for YAGO. As mentioned above, there are more than 100,000 types in YAGO. Obviously, it was not efficient and unwise to populate pairwise disjointness assertions on them manually. Instead, the assertions were added for high-level types and an automated reasoning approach was used to populate those assertions for the rest of the types.

The second Knowledge Graph extension used in DeepQA can be viewed as predefined graph patterns for selected entity types. More precisely, it was very similar to the entity description patterns, which were defined in Sect. 6.2. In Jeopardy!, questions were usually asked about a certain aspect of an entity with clues provided on its other aspects. For example, a representative question of such type could be "Bram Stoker' s famous book on vampires was published in this year." The entity type in this example is *Book* whose aspects (predefined attributes) are *author* (Bram Stocker), *topic* (vampires) and *publish year* (*asked aspect*). Basically, once a question can be mapped to such a star-shaped graph pattern (called as *Frame* in Watson DeepQA), it would be more efficient and accurate to get its answer. To achieve a better performance in final Jeopardy!, the team built frames for U.S. presidents and vice presidents, U.S. states and their capitals, countries and their capitals, books, movies and awards (such as Oscar, Nobel etc.).

7.3.3 How Knowledge Graphs Are Used in Watson DeepQA?

Watson DeepQA's core functionality is to analyse huge amounts of unstructured data, consisting of documents written in the natural language, and to answer a wide variety of questions spanning a broad range of topics. However, the structured data coming from knowledge sources played a crucial role in making Watson successful. In fact, even if the structured data typically cover narrow domains of knowledge, they allow formal semantics and logical reasoning techniques that are able to provide very precise evidences and take into account implicit knowledge. Databases, knowledge bases and ontologies support Watson both in generating candidate answers and in finding supporting evidences by complementing the Watson's ability to work with any natural language statement with the use of the background knowledge provided by structured data. Watson DeepQA used the above-mentioned knowledge graphs in many ways. They have proven able to improve performance and overcome limitations of hybrid approaches based on other techniques. Specifically, knowledge graphs were used for temporal and spatial reasoning, detecting and scoring temporal/spatial relations, and computing the type coercion (TyCor). We give here a brief description of the two processes. Interested readers can refer to [127] for detailed discussions. Watson DeepQA used the above-mentioned knowledge graphs in many ways. They have proven able to improve performance and overcome limitations of hybrid approaches based on other techniques. Specifically, knowledge graphs were used for temporal and spatial reasoning, detecting and scoring temporal/spatial relations and computing the YAGO type coercion (TyCor). Here, a brief description of the two processes is given. Interested readers can refer to [127] for detailed discussions.

Using Knowledge Graph for Temporal and Geospatial Reasoning

Watson DeepQA can be considered an *evidence-based* system, which means that evidences were gathered for supporting or refuting candidate answers. These evidences are eventually combined to rank candidate answers. Among these evidences, the geospatial and/or temporal relations are one of those features that require knowledge graph data and techniques.

To calculate temporal relations, questions are first analysed to detect temporal expressions. On detection, such expressions are syntactically analysed and given a logical form (clause). Then, candidate answers supplemented with temporal clauses are checked against Knowledge Graphs. For each candidate, the process results in evaluating whether the available knowledge supports the candidate or not.

For example, consider the question: "*At the museum you can see the spinal column of this 19th century Presidential assassin.*" The analysis step will identify the temporal expression: "*this 19th century Presidential assassin,*" which is converted into the conjunctive clause like *livedIn*(?*x*, 19th century), *Presidential_ Assassin*(?*x*). Such queries will be executed against a knowledge graph that contains the facts of $Born(George\_Wallace, 08/25/1930)$, $Born(John\_Booth, 01/25/1838)$, $Lawyer(George\_Wallace)$, $Murderer(John\_Booth)$. By checking the temporal compliance, the system can rule out *George_Wallace* because he was not born in the nineteenth century.

Many questions that Watson DeepQA needed to answer also required one to understand the geospatial relations between the question and both potential answers. Knowledge graphs are employed by the system to determine whether candidate answers have the same geospatial relation as that expressed in the question. For example, the trivia *"THE HOLE TRUTH (1200): Asian location where a notoriously horrible event took place on the night of June 20, 1756"* whose answer is the *"Black Hole of Calcutta"* has abundant textual evidences mentioning either India or Calcutta but not explicitly stating that it is in Asia. Linked Data like Geo Names allow the Watson to infer this relation. Evidences provided both by temporal and geospatial relations are scored and contribute to support or refute candidate answers.

Computing Type Coercion

One of the most important problems that Watson DeepQA has to solve is the type coercion (TyCor). TyCor problem consists in checking wherever a particular candidate answer's type matches the one (see LAT in Sect. 7.3.1) of the question. Because language does not distinguish between instantiation (e.g. Calcutta is a city) and subclassing (e.g. Indian city is a subclass of Asian city), solving this problem requires to address two important aspects: Entity Disambiguation and Matching (EDM) [172] to map textual mentions of entities with proper knowledge-base resources and Predicate Disambiguation and Matching (PDM) to map the LAT of the question to a concept in the ontology.

In Watson DeepQA serveral algorithms have been used to accomplish the TyCor task, one of those named *YAGO TyCor* [126] performs taxonomic reasoning and is based on DBpedia, YAGO and other sources. Taxonomic reasoning uses structured data in four steps: it finds entities in the source (such as DBpedia) that correspond to the candidate answer, taking into account both synonymy and polisemy; it retrieves the types for each entity; identifies types that correspond to the expected LAT; and finally it evaluates the degree of match between types related to the candidate answers with types related to LAT. Reasoning tasks, such as including subsumption and disjointness computation, and instance checking, are performed in every step.

7.3.4 Lessons Learnt from Watson DeepQA

Here we summarise the lessons learnt from Waston DeepQA:

- **Using both structured and unstructured resources** Watson DeepQA is different from those QA systems that rely on pure structured knowledge. DeepQA can make use of both formal structured knowledge and unstructured resources.
- **Mixing statistical methods with knowledge-based logical reasoning is a key move** Watson DeepQA employs both statistical methods and logical methods. The former use unstructured sources and NLP techniques and can provide broad coverage but are less precise; the latter use formal structured knowledge covering a narrower range of questions but are more precise. So they are natural complements to each other.
- **Good training data guarantees the performance of learning methods** Good training data is required to tune up the learning machinery.
- **QA over textual sources is feasible under parallel working** Open-domain QA over textual sources is feasible, although it is a non-trivial task. Even if, in general, the problem is computationally challenging, parallelising QA pipelines allows scalability to a certain extent.

- **Knowledge Graph finds semantic evidence** Knowledge acquisition and curation is key to achieving good results, this is especially true for domain-specific QA. In DeepQA knowledge graph can find semantic evidence for *temporal* and/or *spatial* constraints in questions. The well-defined semantics of temporal and spatial evidence is particularly useful for explanatory purposes.
- **Lightweight semantics** Shallow temporal and spatial semantics are used in DeepQA so that it only employ light weight reasoning efforts.

7.4 Using Knowledge Graphs for Improving Textual Question Answering*

This section describes a flexible and adaptable architecture for the design of advanced QA systems, which the University of Trento had developed in the last decade [37, 67, 166–170, 208–212], also thanks to the collaboration with IBM Watson Research Center, NY. This architecture illustrates how knowledge graphs can be used in the question answering pipeline and, more interestingly, it shows performance differences among various knowledge graphs.

Specifically, this section starts with a brief introduction of a NLP pipeline well suited for working with knowledge graphs. Then, it will put special emphasis on the re-ranking module. The re-ranker is based on the structural kernel technology, which enables the easy encoding of trees and graphs in learning algorithms, such as SVMs. This way, it could exploit two different types of graph information: (i) syntactic and semantic graphs derived by the syntactic and semantic analysis of the question and answer passage pairs; and (ii) external Knowledge Graphs such as DBpedia, which we exploit for enriching the previous graph with semantic and typed information.

This section also shows the results of different comparative experiments performed with the system. They aim at measuring the influence of using different knowledge graphs within various re-ranking models.

7.4.1 A Flexible QA Pipeline

The proposed pipeline takes in input a question, a document corpus and some machine learned models to output a set of possible answers. It is based on the UIMA framework and has both an interactive and an experiment mode. The pipeline includes a set of natural language analysis components, a mechanism for caching results of computationally expensive analysis, a module for Machine Learning based re-ranking, a question classifier matched with NERs (named entity recognition tasks), a question focus identifier, and an evaluation module.

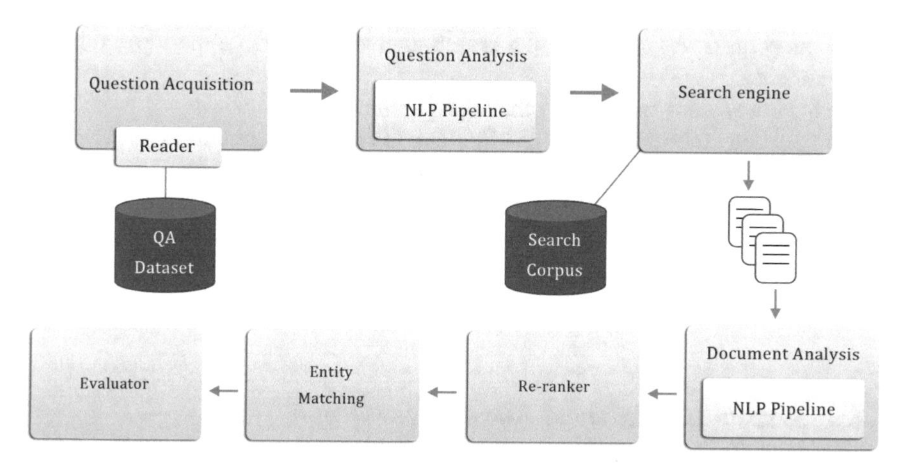

Fig. 7.2 Flexible question answering architecture

The Question Acquisition module has two different variants: the first is used in the interactive version of the pipeline and takes the question from the user input; the other reads the questions, the answers and their associations from a QA dataset stored on the disk and it is used in the version of the pipeline for carrying out experiments. In the second case, the module can be specialised for reading specific QA datasets. A QA dataset may contain questions and answers represented in an XML format and another one in CSV. When there are differences in how the information is represented, specific readers who send the data to the QA Data module can be developed. The QA Data module is responsible for managing the information needed for evaluating the system after the entire processing. The Question Analysis module performs all the NLP tasks on the questions. The Retrieval module uses a search engine to retrieve the documents relevant to the question. The generated query contains the terms of the question. The Document Analysis module performs NLP analyses on documents. The CACHE component stores the information which are expensive to compute and makes them available for the next pipeline executions. The Re-ranking module performs the re-ranking of the paragraphs retrieved by the Retrieval module, using a given model. The Entity Matching module takes the output of the question classifier and checks if a document contains entities of a type matching the question type. The last two modules produce a permutation of the documents list and an improvement of the ranking. The Evaluation module uses the data from the QA Data module to compute the evaluation measures. In the interactive version, the system does not output the evaluation results but shows the first ranked passage or the first entity matching the question type (Fig. 7.2).

The steps executed by the systems are the following:

1. the questions and the data for evaluation are read from the dataset; in the interactive version, a question is read from the user input;
2. the question is analysed by the NLP pipeline;

3. the question is used to query the search engine and retrieve the top-k relevant documents;
4. the documents are analysed by the NLP pipeline;
5. the question classifier output and the named entity recognition tasks' (NERs') output are used to reorder the retrieved document list;
6. the re-ranker is used to produce a permutation of the retrieved document list;
7. the evaluation of the document list is carried out and the results are averaged for all the questions. In the interactive version, the entity matching the question type is returned.

7.4.2 Exploiting External Knowledge (Graphs) for Re-ranking

The Re-ranking module uses a model produced by an SVM (support vector machines) learner to perform classification and reorder the paragraphs list returned by the search engine. The re-ranker performance depends on how the model is engineered. In the case of QA systems, the question and the paragraphs which may contain the answer form question and answer pairs. A pair can be put in relation to another, and the machine can learn which pair is more likely to be a valid question and answer pair. In a labelled training set, there will be examples constituted by pairs of pairs. If the first pair encodes a valid question and answer the example will be positive. If the valid answer is in the second pair, the example will be negative. Of the two pairs, one must contain a valid answer and the other must not. The question and the answer in a pair can be represented in different ways and with different features, depending on how the model is designed. The training data is used to build a re-ranking model. At re-ranking time the model is loaded and the classifier is applied to unseen examples, namely the current question paired with one of the paragraphs to re-rank. This data forms the first pair of the example. The second pair is empty. In this way, the classifier will output its confidence about the paragraphs containing a valid answer for the question. The paragraphs can eventually be sorted by the correspondent confidence values, producing a new ranking.

As stated for the previous Machine-Learning tasks, re-ranking requires two steps: learning and classification.

The learning phase requires that each question in the training set comes with the list of paragraphs retrieved by the search engine for that question. The search engine ranking score of each paragraph is kept for later use. Information about which paragraphs contain a valid answer is also required. This data is sufficient for generating the learning examples. A question is paired with each paragraph in the list. The pairs containing a valid answer form a group, while the remaining form the other. A Cartesian product between the former and the latter produces pairs with the valid answer in the first question and answer pair. The product result can be used to generate positive and negative examples in alternation, swapping the pairs in case of negative examples.

Figure 7.3 can be observed to get a clearer idea of the generation process.

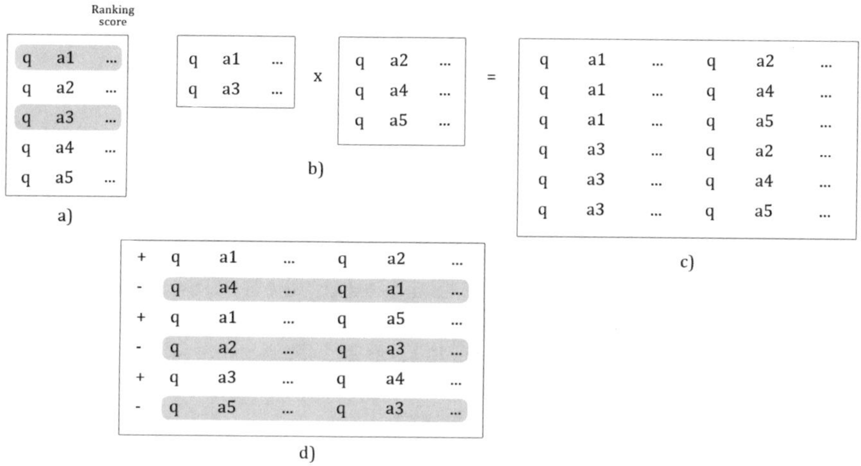

Fig. 7.3 Training data generation

The (a) box contains a question and the candidate answers, namely the passages, retrieved by the search engine. The ranking scores of candidate answers are also included. Obviously, the question is the same for each candidate answer. The highlighted rows contain candidates which are good answers for the question. In (b) the correct answers form a group, and the wrong ones another. The two groups are multiplied by a Cartesian product. The result is given in the (c) box, where there are two question answer pairs on each row. All the pairs of pairs in the box constitute positive examples for the training, where the left pair is the good one and the right pair is the bad one. Negative examples are generated in alternation with positive examples. As displayed in the (d) box, the positive examples are on the odd rows and the negative, highlighted, are on the even rows. The difference between the (c) box and the (d) box is that in the negative examples the pairs are swapped, since it is not acceptable that a wrong answer is placed on the left, and thus considered better than a good answer. Using this generation mechanism, the positive and negative examples are balanced.

The QA Data module is used to generate the training examples, which are stored in a file. In the examples, questions and answers are represented with their constituency trees, and as additional features there are the search engine ranking scores of the answers. The content of the file can be seen as a basic representation of the data needed for the re-ranker to learn a re-ranking model.

In the following sections, all the models implemented are described. Each of them includes the search engine ranking score of a given paragraph. The CH, CH+QC+TFC and CH+QC+TFC+TM models are presented in [208, 211, 234]. The latter has the best performance. The figures associated with the models contain the representation of the sample question *"What company owns the soft drink brand "Gatorade""* on the left, and the answer *"Stokely-Van Camp bought the formula and started marketing the drink as Gatorade in 1967. Quaker Oats Co. took over Stokely-Van Camp in 1983."* on the right. The representations are simplified by keeping only the salient words and are intended to convey the characteristics of the models.

Shallow Chunk Tree (CH)

In the Shallow Chunk Tree model, questions and paragraphs are encoded into two trees, where lemmas constitute the leaf level, the part-of-speech (POS) tags are at the pre-terminal level and the sequences of POS tags are organised into the third level of chunk nodes. We encoded structural relations using the **REL** tag, which links the related structures in the question and the candidate answer paragraph, when there is a match between the lemmas in the question and the paragraph. We marked the parent (POS tags) and grand parent (chunk) nodes of such lemmas by prepending a **REL** tag. Figure 7.4 illustrates the CH model.

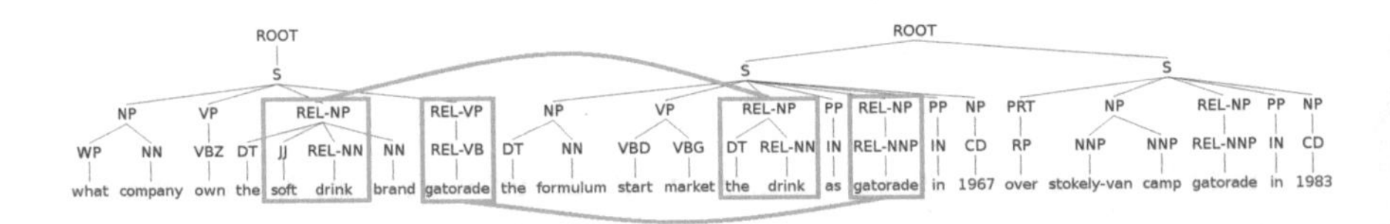

Fig. 7.4 CH model

We also prune the trees by removing the nodes beyond a certain distance (in terms of chunk nodes) from the **REL** and *REL-FOCUS* nodes. This removes irrelevant information and speeds up learning and classification.

Using Wikipedia for REL Matching (wiki)

The lemma matching strategy employed in **CH** for detecting **REL** structures may result in low coverage, e.g. it is not able to match different variants for the same name. We partially remedy this by using the wikification tools. They recognise lemmas that may denote Wikipedia pages in plain text and disambiguate them to obtain a unique Wikipedia page. Such tools determine whether a certain lemma may denote a Wikipedia page(s) by looking it up in a precomputed vocabulary created using Wikipedia page titles and internal link network [58, 160].

If two word sequences in a question and in an answer, respectively, have been annotated with the same Wikipedia link we consider them as matching and add new **REL** tags to the question and answer passage representations.

Table 7.1 Question class/named entity types compatibilities [211]

Question category	Named entity types
HUM	Person
LOC	Location
NUM	Date, time, money, percentage
ENTY	Organisation, person

CH + Question Classification + Focus Detection (CH+QC+TFC)

This model is similar to the CH model, except for additional semantic information provided by the **Question Type**, **NER** and **Question Focus** annotators. We encode information about the question focus and availability of the named entities compatible with the question type in the answer by using the **REL-FOCUS-<QC>** tag, where **<QC>** is substituted with the question type tag. We mark (i) the focus chunk in the question and (ii) the answer passage chunks containing named entities of type compatible with the question class, by prepending the above tags to their labels. The compatibility between the categories of named entities and questions is evaluated with a lookup to a manually predefined mapping presented in Table 7.1. Figure 7.5 illustrates an example of such structure.

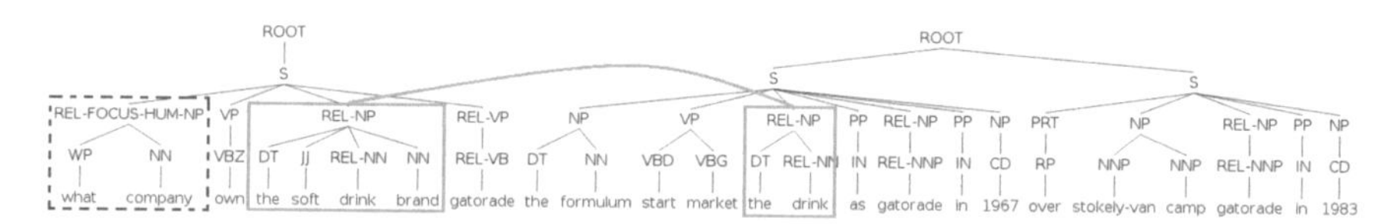

Fig. 7.5 CH+QC+TFC model

CH + Question Classification + Focus Detection* (CH+QC+TFC*)

This model is the same as above, except for the fact that we slightly modify the **REL-FOCUS** encoding into the tree. Instead of prepending **REL-FOCUS-<QC>**, we only prepend **REL-FOCUS** to the target chunk node, and add a new node *QC* as the rightmost child of the chunk node, e.g. in Fig. 7.5, the focus node would be marked as **REL-FOCUS** and the sequence of its children would be *[WP NN HUM]*. This modification intends to reduce the feature sparsity.

CH + Question Classification + Focus Detection + Type Match Information (CH+QC+TFC+TM)

This model incorporates information about *subclass of* and *is instance of class* relations between entities and classes mentioned in the question and answer passages.

We say that two lemmas are in the *type match (TM)* relation if they refer to two entities or an entity and a class from an external knowledge graph linked by the *rdfs:subclassOf* or *rdf:type* relation. This pipeline employs YAGO [116], DBpedia [33] and WordNet [74].

More formally, given two lemmas or lemma sequences, a_1 and a_2, belonging to two text passages, p_1 and p_2, respectively, and given an $R(a, p)$ function, which returns an ID of a LD class/entity corresponding to a in passage p, we define *TM* (r_1, r_2) as

$$\begin{array}{c} isa\,(r_1, r_2) : if\ isEntity\,(r_1)) \wedge isClass\,(r_2) \\ subClassOf\,(r_1, r_2) : if\ isClass\,(r_1) \wedge isClass\,(r_2) \end{array} \tag{7.1}$$

where $r_1 = R(a_1, p_1)$, $r_2 = R(a_2, p_2)$ and *isEntity(r)* and *isClass(r)* return true if r is an entity or a class, respectively, and *false* otherwise. It should be noted that, due to the ambiguity of natural language, the same anchor may have different references depending on the context.

We denote the set of preterminal parents of lemmas in TM relation as N_{TM}. We have considered the following strategies of encoding TM relation in the parse trees: (i) **TM node** (TM_N). Add leaf sibling tagged with *TM* to all the nodes in N_{TM}. (ii) **Directed TM node** (TM_{ND}). Add leaf sibling tagged with **TM-CHILD** to all the nodes in N_{TM} corresponding to the subclass/entity, and leaf siblings tagged with **TM-PARENT** to the nodes corresponding to the class. (iii) **Focus** *TM* (TM_{NF}). Add leaf siblings to all the nodes in N_{TM}. If *matchedTokens* is a part of a question focus label them as **TM-FOCUS**. Otherwise, label them as **TM**. (iv) **Combo** TM_{NDF}. Encode using the TM_{ND} strategy. If *matchedTokens* is a part of a question focus label then also add a child labelled *FOCUS* to each of the TM labels. Intuitively, TM_{ND}, TM_{NF}, TM_{NDF} are likely to result in more expressive patterns. Figure 7.6 shows an example of the TM_{ND} annotation.

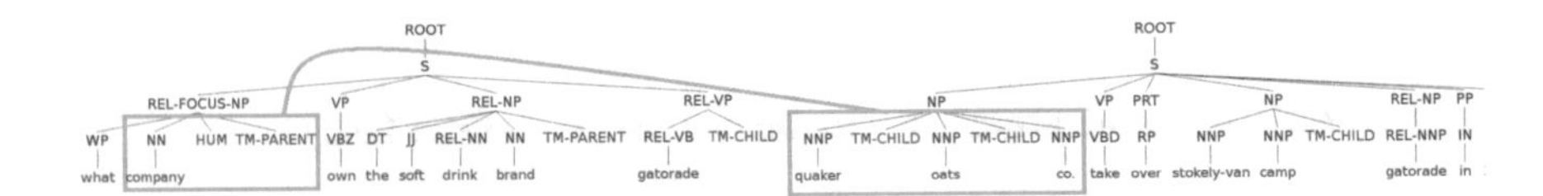

Fig. 7.6 CH+QC+TFC+TM models

SVM Classifier

The SVM classifier available for components in the pipelines is the SVMLightTK 1.5 classifier. SVMLightTK[6] is our extension of Thorsten Joachims' SVMLight.[7] It adds the support for feature vectors, tree kernels, sequences, forest and set of vectors.

[6]http://disi.unitn.it/moschitti/Tree-Kernel.htm.

[7]http://svmlight.joachims.org/.

The classifier is available as a dynamic library. A Java Native Interface exposes the methods needed to instantiate it and classify the examples. The software is integrated and the end user is only required to instantiate a wrapper class, load a model and perform classification.

7.4.3 Evaluation: Impact of Knowledge Graphs in Semantic Structures

Dataset. The system performance is evaluated on the Text REtrieval Conference (TREC) datasets. TREC was started in 1992 and it has several tracks, areas of focus in which particular retrieval tasks are defined. The Question Answering track started in 1999 and every year the complexity of the challenge increases. The corpus employed for testing is the AQUAINT-1 corpus, used also in the tracks from 2002 to 2006. The formulation and focus of these tracks vary. Questions are proposed in different ways and the only tracks having a format suitable for a system answering to factoid question are the 2002 and 2003 tracks. Indeed, in these tracks datasets contain complete factoid questions with the corresponding answers, while the others contain questions asking for list of entities, or a set of questions without a subject which is stated before them. The employed datasets contain 1403 questions. After the removal of questions not having associated answers, they become 824.

We run the experiments in fivefold cross-validation. This means that we split our questions into five non-overlapping subsets, s_i $(i = 1, 5)$. Then iteratively, with i ranging from 1 to 5, we use the union of four subsets, $\bigcup_{j=1,5;j\neq i} s_j$, for training, and predict on s_i. We retrieve 10 and 50 candidate answers per question when training and testing, respectively.

Feature Vectors

We used a subset of the similarity functions between question (Q) and answer passage (AP) described. These are used along with the structural models. More explicitly: *Term-overlap features*: i.e. a cosine similarity over question/answer, $sim_{COS}(Q, AP)$, where the input vectors are composed of lemma or POS-tag n-grams with $n = 1, .., 4$. *PTK score*: i.e. output of the Partial Tree Kernel (PTK), defined in [165], when applied to the structural representations of Q and AP, $sim_{PTK}(Q, AP) = PTK(Q, AP)$ (note that, this is computed within a pair). PTK defines similarity in terms of the number of substructures shared by two trees. *Search engine ranking score:* the ranking score of our search engine assigned to AP divided by a normalising factor.

Search engines. We adopted Terrier[8] using the accurate BM25 scoring model with default parameters. We trained it on the TREC corpus (3Gb), containing about 1 million documents. We performed indexing at the paragraph level by splitting each document into a set of paragraphs, which are then added to the search index. We retrieve a list of 50 candidate answer passages for each question.

---

[8]http://terrier.org.

Wikipedia link annotators. We use the Wikipedia Miner (WM) [160][9] tool and the Machine Linking (ML)[10] Web services to annotate Q/AP pairs with links to Wikipedia. Both tools output annotation confidence. We use all WM and ML annotations with confidence exceeding 0.2 and 0.05, respectively. We obtained these figures heuristically, they are low because we aimed to maximise the Recall of the Wikipedia link annotators in order to maximise the number of TMs. In all the experiments, we used a union of the sets of the annotations provided by WM and ML.

Results

These experiments evaluated the accuracy of the following models (described in the previous sections): (i) a system using Wikipedia to establish the REL links and (ii) systems that use knowledge graphs to find type matches (TM).

We experimented with various LOD Knowledge Repository combinations and report the best results in Table 7.2.

The first header line of the Table 7.2 shows which baseline system was enriched with the TM knowledge. *Type* column reports the TM encoding strategy employed (see Sect. 7.4.2). *Dataset* column reports which knowledge source was employed to find TM relations. Here, *yago* is YAGO2,[11] *db* is DBpedia, and *wn* is WordNet 3.0. The first result line in Table 7.2 reports the performance of the strong **CH+V** and **CH+V+QC+TFC*** baseline systems. Line with the "wiki" dataset reports on **CH+V** and **CH+V+QC+TFC*** using both Wikipedia link annotations provided by ML and MW and hard lemma matching to find the related structures to be marked by **REL**. The remainder of the systems is built on top of the baselines using both hard lemma and Wikipedia-based matching.

The tables show that the systems exploiting LOD knowledge outperform the strong **CH+V** and **CH+V+QC+TFC*** baselines. Note that **CH+V** enriched with TM tags performs comparably to, and in some cases even outperforms, **CH+V+QC+TFC***. Compare, for example, the outputs of **CH+V+**TM_{NDF} using YAGO, WordNet and DBpedia knowledge and those of **CH+V + QC + TFC*** with no LOD knowledge.

Adding TM tags to the top-performing baseline system, **CH+V+QC+ TFC***, typically results in a further increase in the performance. The best-performing system in terms of MRR and P@1 is **CH+V+QC+TFC*+**TM_{NF} system using the combination of WordNet and YAGO2 as a source of TM knowledge and Wikipedia for REL-matching. It outperforms the **CH+V+QC+TFC*** baseline by 3.82 and 4.15 % in terms of MRR and P@1, respectively. Regarding MAP, a number of systems employing YAGO2 in combination with WordNet and Wikipedia-based REL-matching obtain 0.37 MAP score thus outperforming the **CH+V+QC+TFC*** baseline by 4 %.

We used the paired two-tailed t-test for evaluating the statistical significance of the results reported in Table 7.2. ‡ and † correspond to the significance levels of 0.05

[9]http://sourceforge.net/projects/wikipedia-miner/files/wikipedia-miner/wikipedia-miner_1.1, we use only a topic detector module that detects and disambiguates anchors.

[10]http://www.machinelinking.com/wp.

[11]http://www.mpi-inf.mpg.de/yago-naga/yago1_yago2/download/yago2/yago2core_20120109.rdfs.7z.

Table 7.2 Results in fivefold cross-validation on TREC QA corpus

Type	Dataset	CH + V			CH + V + QC + TFC*		
		MRR	MAP	P@1	MRR	MAP	P@1
–	None	36.82 ± 2.68	0.30 ± 0.02	26.34 ± 2.17	40.50 ± 2.32	0.33 ± 0.02	31.46 ± 2.42
–	wiki	39.17 ± 1.29‡	0.31 ± 0.01‡	28.66 ± 1.43‡	41.33 ± 1.17	0.34 ± 0.01	31.46 ± 1.40
TM_N	yago	40.71 ± 2.07	0.33 ± 0.03†	30.24 ± 2.09‡	43.28 ± 1.91†	**0.36 ± 0.01**†	33.90 ± 2.75
TM_N	yago+wn	**42.01 ± 2.26**‡	**0.34 ± 0.02**‡	**32.07 ± 3.04**‡	**43.98 ± 1.08**‡	**0.36 ± 0.01**‡	**35.24 ± 1.46**‡
TM_N	yago+wn+db	41.52 ± 1.85‡	**0.34 ± 0.02**‡	30.98 ± 2.71‡	43.13 ± 1.38	**0.36 ± 0.01**	33.66 ± 2.77
TM_{NF}	yago	**42.01 ± 2.44**†	**0.34 ± 0.02**‡	**32.07 ± 3.01**‡	43.82 ± 2.36†	0.36 ± 0.02‡	34.88 ± 3.35
TM_{NF}	yago+wn	41.69 ± 1.66‡	**0.34 ± 0.02**‡	31.10 ± 2.44‡	**44.32 ± 0.70**‡	0.36 ± 0.01‡	**35.61 ± 1.11**‡
TM_{NF}	yago+wn+db	41.56 ± 1.41‡	**0.34 ± 0.02**‡	30.85 ± 2.22†	43.79 ± 0.73‡	**0.37 ± 0.01**†	34.88 ± 1.69‡
TM_{ND}	yago	42.11 ± 3.24‡	0.34 ± 0.02‡	32.07 ± 4.06†	44.04 ± 2.05‡	0.36 ± 0.01‡	34.63 ± 2.17‡
TM_{ND}	yago+wn	**42.96 ± 1.45**‡	**0.35 ± 0.01**‡	**33.05 ± 2.04**‡	**44.25 ± 1.32**‡	**0.37 ± 0.00**‡	**34.76 ± 1.61**‡
TM_{ND}	yago+wn+db	42.56 ± 1.25‡	**0.35 ± 0.01**‡	32.56 ± 1.91‡	43.91 ± 1.01‡	**0.37 ± 0.01**†	34.63 ± 1.32‡
TM_{NDF}	yago	42.31 ± 2.57‡	**0.35 ± 0.02**‡	32.68 ± 3.01‡	**44.22 ± 2.38**‡	**0.37 ± 0.02**‡	**35.00 ± 2.88**‡
TM_{NDF}	yago+wn	42.80 ± 1.19‡	**0.35 ± 0.01**‡	33.17 ± 1.86‡	43.91 ± 0.98‡	**0.37 ± 0.01**‡	34.63 ± 0.90‡
TM_{NDF}	yago+wn+db	**43.15 ± 0.93**‡	**0.35 ± 0.01**‡	**33.78 ± 1.59**‡	43.96 ± 0.94‡	**0.37 ± 0.01**‡	34.88 ± 1.69‡

and 0.1, respectively. We compared (i) the results in the *wiki* line to those in the *none* line and (ii) the results for the *TM* systems to those in the *wiki* line.

We typically obtain better results when using YAGO2 and/or WordNet. In our intuition this is due to the fact that these resources are large-scale, have fine-grained class taxonomy and contain many synonymous labels per class/entity thus allowing us to have a good coverage with TM links. DBpedia ontology that we employed in the *db* experiments is more shallow and contains fewer labels for classes, therefore the amount of discovered TM matches is not always sufficient for increasing performance. YAGO2 provides a better coverage for TM relations between entities and their classes, while WordNet contains more relations between classes.[12]

Different TM-knowledge encoding strategies, TM_N, TM_{ND}, TM_{NF}, TM_{NDF}, produce small changes in accuracy. We believe that the difference between them would become more significant when experimenting with larger corpora.

7.4.4 Conclusion

This section has given an overview of the possible usages of Knowledge Graphs in state-of-the-art QA systems. For NLP-based systems, with reference to structural kernel frameworks, it has been shown how to improve re-ranking algorithms by encoding graph information of two different types:

- Syntactic/Semantic graphs generated from the question and answer passage text; and
- Linked Open Data semantic graph information, again introduced in the above graph in terms of the similarity between sentence constituents.

The benefit of the adoption of Knowledge Graphs as external knowledge for helping re-ranking methods has been evaluated using TREC benchmarks. Interestingly, thanks to a suitable use of knowledge sources, such as Wikipedia, relatively simple frameworks such as those described in Sect. 7.4.1 can approximate to some extent the accuracy of more complete state-of-the-art systems like IBM Watson.

Future research directions regarding the use of explicit knowledge resources in QA include:

1. the development of new re-ranking models including new engineered features or rules;
2. experiments with bigger training datasets (from 100k up to 1 million examples). As shown in [208], this can lead to a large improvement; and
3. more advanced representation of question and answer passage pairs, for example, based on dependency structures, semantic role labelling and discourse structure.

[12]We consider the WordNet synsets to be classes in the scope of our experiments.

Part III
Industrial Applications and Successful Stories

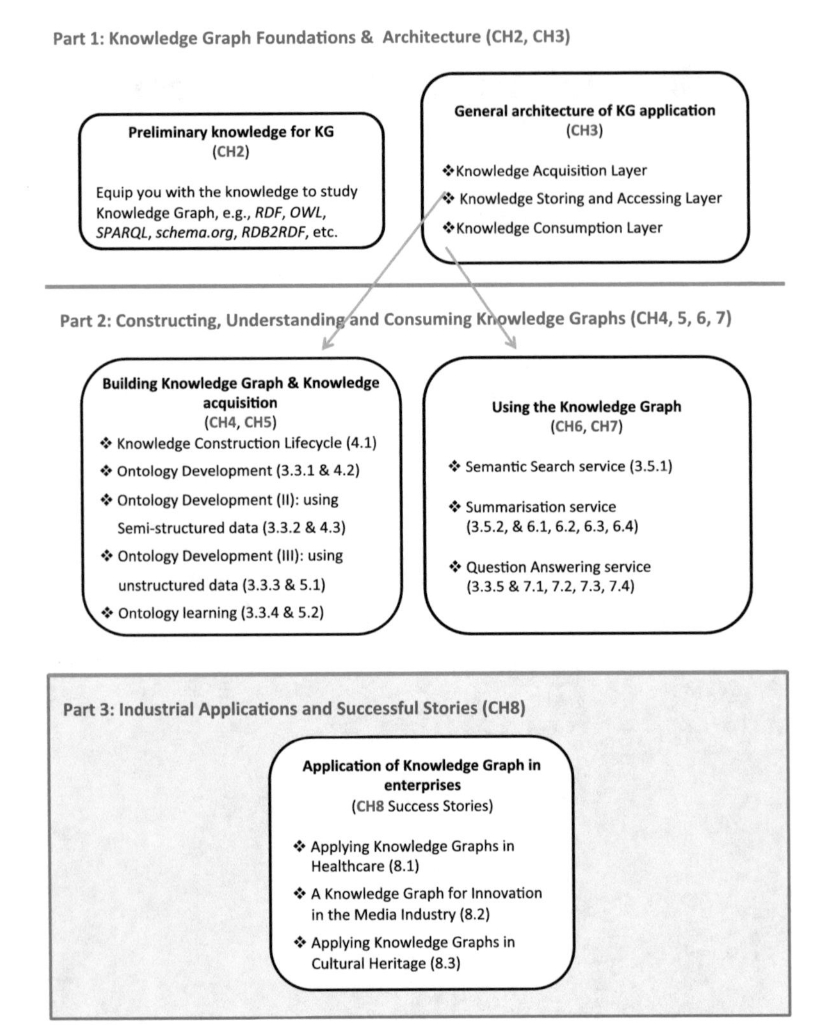

Fig. RoadMap. 3 The roadmap of Part III

In this part we introduce some selected applications of knowledge graph techniques on various domains: (i) media (Sect. 8.1), (ii) culture (Sect. 8.2), (iii) healthcare (Sect. 8.3).

Part III is structured as follows:

Chapter 8: Success Stories
Section 8.1: A Knowledge Graph for Innovation in the Media Industry
Section 8.2: Applying Knowledge Graphs in Cultural Heritage
Section 8.3: Applying Knowledge Graphs in Healthcare

Chapter 8
Success Stories

Marco Monti, Fernanda Perego, Yuting Zhao, Guido Vetere, Jose Manuel Gomez-Perez, Panos Alexopoulos, Hai Nguyen, Gemma Webster, Boris Villazon-Terrazas, Nuria Garcia-Santa and Jeff Z. Pan

So far, we have introduced different approaches to construct, explore and exploit knowledge graphs in large organisations. However, it is also necessary to explore the practical applications of knowledge graphs and their related technologies in real-life scenarios, focusing on the associated benefits and potential drawbacks and reflecting

M. Monti (✉) · F. Perego · Y. Zhao
IBM Italia, Circonvallazione Idroscalo, 20090 Milan, Italy
e-mail: marco.monti@it.ibm.com

F. Perego
e-mail: fperego@it.ibm.com

Y. Zhao
e-mail: yuting.zhao@it.ibm.com

G. Vetere
IBM Italia, via Sciangai 53, 00144 Rome, Italy
e-mail: gvetere@it.ibm.com

J.M. Gomez-Perez · P. Alexopoulos · G. Webster · B. Villazon-Terrazas · N. Garcia-Santa
Expert System, Prof. Waksman 10, 28036 Madrid, Spain
e-mail: jmgomez@expertsystem.com

P. Alexopoulos
e-mail: palexopoulos@expertsystem.com

G. Webster
e-mail: gwebster@abdn.ac.uk

B. Villazon-Terrazas
e-mail: bvillazon@expertsystem.com

N. Garcia-Santa
e-mail: ngarcia@expertsystem.com

H. Nguyen · J.Z. Pan
University of Aberdeen, King's College, Aberdeen AB24 3UE, UK
e-mail: hai.nguyen@abdn.ac.uk

J.Z. Pan
e-mail: jeff.z.pan@abdn.ac.uk

J.Z. Pan et al. (eds.), *Exploiting Linked Data and Knowledge Graphs in Large Organizations*, DOI 10.1007/978-3-319-45654-6_8

on the lessons learnt. In this chapter, we elaborate on some of the cases introduced in Chap. 1 and present new success stories, providing a detailed account of the application of knowledge graph methods and technologies in different organisations and domains. To this purpose, we found the domains of media, culture and healthcare to be particularly compelling. Furthermore, the selected success stories cover a broad spectrum of potential adopters of knowledge graph technologies, including large IT providers, multinational companies and educational organisations.

8.1 A Knowledge Graph for Innovation in the Media Industry

8.1.1 The Business Problem

The communication between brands and consumers is set to explode. Product features are no longer the key to sales and the combination of both personal and collective benefits is becoming an increasingly crucial aspect. As a matter of fact, brands providing such value achieve a higher impact and consequently derive clearer economic benefits. On the other hand, millennials are taking over; inducing a dramatic change in the way consumers and brands engage and what channels and technologies are required to enable the process. As a result, traditional boundaries within the media industry are being stretched and new ideas, inventions and technologies are needed to keep up with the challenges raised by the increasing demands of this data-intensive, in-time, personalised and thriving market.

Thus, it is necessary to leverage advances in the area by stimulating a collaboration ecosystem between the different players. Inspiring examples include the adoption by Tesla Motors of the open patent policy, whereby Tesla shares their innovation in regard to electric cars openly via the Internet. In return, Tesla expects the industry to further evolve the electric car and dynamise the market. In the media industry, a paradigmatic case of this *better together* approach is HAVAS 18 Innovation Labs, deployed at strategic locations around the world. One such location is the Siliwood research centre in Santa Monica, co-created with Orange, which focuses on the convergence between technology, data science, content and media. The 18 Innovation Labs seeks to connect a great mix of local talent over the sites, involving innovators, universities, start-ups and technology trends to co-create initiatives relevant now and in the midterm for both HAVAS and their clients to stay one step ahead.

To achieve that, HAVAS has created an enterprise knowledge graph and information platform that aggregates all the available knowledge about technology start-ups worldwide and makes it available for exploitation by media business strategists through a single entry point. To the best of our knowledge this is one of the first applications of knowledge graph principles in the enterprise world, and the first in the media industry, after Internet search giants Google, Yahoo! and Bing coined the term at Web-scale, each with their own implementations. Related initiatives include domain-specific efforts such as social graphs like Facebook, and reference resources like DBPedia and Freebase.

8.1.2 The HAVAS 18 Knowledge Graph

In a way analogous to the above-mentioned initiatives, the main objective of the HAVAS 18 knowledge graph is to enable knowledge-based services for search, discovery and understanding of information about relevant start-ups in their first 18 months. So, we aimed at providing a unified knowledge graph where:

1. Entities are uniquely identified by URIs and interlinked across sources.
2. Such entities are relevant to HAVAS 18 Labs, including start-ups, people of interest related to them through different roles, e.g. founder, investor, etc. bigger and more established companies, universities and technology trends.
3. Rich information is provided about entities (facts, relationships and features).

To this purpose, the graph follows the lifecycle described in Chap. 4, comprising three main phases: knowledge acquisition, integration and consumption. We extract data from online sources, including generalist and specialised Web sites, online news, entrepreneurial and general-purpose social networks, and other content providers. By maximising the use of sources offering Web APIs, we expect to minimise additional unstructured data processing time and complexity, at the cost of unexpected changes in the APIs, a potential source of decay in the knowledge graph. Data sources include:

- **Core data** from specialised sites like AngelList and CrunchBase, with useful facts about the main entities in the graph (start-ups, innovators, investors, other companies and universities), the relationships between them, and domain-specific news.
- **Relationships**: Beyond factual knowledge about the entities, the resulting graph lays emphasis on how they are related to each other. We enrich the relationship graph with the information from Facebook, LinkedIn and Twitter which helps in completing a social and professional graph between the entities. Such explicit relationships support the discovery of new insights and navigation.
- **Extended media coverage**, with general news coming from the media in any domain through Newsfeed.ijs.si. News text is processed with the semantic annotation framework Knowledge Tagger (see Sect. 4.4) in order to resolve entities and disambiguate.

The data are structured and integrated in an RDF dataset. The underlying schema is built on top of a number of standard W3C vocabularies, including Schema.org, FOAF and SKOS, and IPTC's (International Press Telecommunications Council) rNews. A service layer is provided on top of the data through a RESTful API with JSON (Java Script Object Notation) and API key authentication. The API allows the exploitation of the graph by application developers and ultimately media business strategists through analytics platforms and dedicated user interfaces. API services include CRUD (Create, read, update and delete) methods for entity and relationship management, graph navigation, search and definition and access to business KPIs (Key Performance Indicators) about the start-ups.

In addition to the means of automated information extraction, the knowledge graph can also be populated with on-site information by local rapporteurs, members of the local entrepreneurial scene distributed at each of the HAVAS 18 Innovation Labs. Rapporteurs are provided with means to add or modify entities and relationships in the graph, following the schema, assisted by autocomplete functionalities that leverage the knowledge previously stored in the graph. They also play the role of curators of knowledge produced either by other peer rapporteurs or extracted automatically. The combination of automatic methods and human expertise allows a common knowledge graph to be leveraged consistently across the company.

Currently, the graph contains information about 1,812 start-ups, 559 technology trends, 1,597 innovators, 20 companies and 35 universities and research centres in Siliwood, following the Linked Data principles. All these entities are additionally connected to relevant online news, where they are mentioned (currently, 36,802), for extended and up-to-date information about them. The Knowledge Graph is updated daily in an automated batch process, identifying new entities and updating existing ones. We expect the Knowledge Graph to quickly reach the threshold of 300,000 start-ups below 18 months and extend to the remaining Labs in the next few months.

8.1.3 Value Proposition

Innovation is often misunderstood and difficult to integrate into companies' mind-set and culture. So, why not activate relevant external talent and resources when necessary? The discovery and surveillance of trends and talent in the start-up ecosystem can be time consuming, though. HAVAS' knowledge graph sets its semantic engineering to run a surveillance monitoring of the entrepreneurial digital footprint, collecting and gathering fruitful insight and information, which provides the Innovation Labs staff with clear leads for their analyses. This is the key approach and philosophy of HAVAS. By automating part of the research process, it can get there faster and more accurately than competitors, leveraging millions of data points and implementing consistency through a single and shared knowledge entry point.

At the moment this assisted process is integrated with a manual audit of trends and start-ups, executing a series of evaluation matrices to weigh and assess each individual entity in the graph against HAVAS' business needs. The knowledge graph is being opened to HAVAS' network, with teams in 120 offices around the world and clients, providing access to knowledge about best-in-class talent to implement new thinking and cutting-edge solutions to the never-ending and evolving challenges within the media industry. Based on the knowledge graph, teams also rate and share experiences, ensuring that learning can be propagated across the network.

8.1.4 Challenges

To optimise the trustworthiness and accuracy of the graph, we maximised the use of authoritative and specialised sources and prioritised freshness over volume. However, entity resolution and disambiguation is an issue, especially when unstructured data from unbounded domains come into play. During data integration and enrichment, several candidate entities can be identified. In order to resolve the correct one we defined evidence models on top of the schema with the key classes required to provide an entity class with univocal context information. For example, in the case of Start-up, this could be Founder, Client and Technology. After data harvesting, e.g. from news, the text is processed by the Knowledge Tagger, which extracts entities and matches them against the evidence model providing a measure of evidence based on the context fragments identified in the text and allowing ranking. Further complexity is added when an entity which is not in the domain of interest, e.g. Domo, the gas station, has to be discriminated from the one which is part of such a domain and potentially in the graph, e.g. Domo, the start-up.

Other challenges include (sub)graph time and version management, reconciliation of automatic versus human updates, and resilience against changes in the data sources, especially Web APIs. It is particularly important to monitor potential decay in a knowledge graph, by applying the existing decay management techniques and methods taken from scientific domains (see [23, 91]), with extant literature on this topic. Once structured as self-contained information packs, personalised subscription, delivery and recommendation of portions of the graph will also be possible.

8.2 Applying Knowledge Graphs in Cultural Heritage

8.2.1 Digital Cultural Heritage and Linked Data

Ever since the first announcement of the Semantic Web vision [28], a number of projects aiming to create open versions of cultural heritage data, including the UK Culture Grid[1] and the Dutch Continuous Access to Cultural Heritage (CATCH) programmes, came into existence. There are also a number of cultural heritage ontologies in existence, including Categories for the Description of Works of Art (J. Paul Getty Trust) and CIDOC CRM (Center for Intercultural Documentation Conceptual Reference Model) [68]. The ontologies and terminologies used are based on a range of technologies, for instance XML and distributed databases as well as RDF/OWL, but there is also an increasing interest in using Semantic Web techniques in this area. For instance, the OpenART [8] project brings an important art research dataset, The

[1] http://www.culturegrid.org.uk.

London Art World 1660–1735 to the Linked Open Data format, so that contents about the art world during that period can be contextualised and linked to the Tate collection and referred to the relevant contemporary artworks. On a larger scale, CultureSampo [154] was developed for publishing heterogeneously Linked Data as a service. The main aim of this project is to create a cultural heritage archive for the whole nation by providing an infrastructure and a set of tools to publish and annotate contents collectively. The case study used in this project is the Finnish cultural heritage archives. The authors argue that semantic linking can add value by facilitating links between artefacts which can lead to a better understanding of the themes or allow the user to make connections more easily. More recently, the work presented in [243] aims to bring library data into the Linked Data world. The authors discuss few limitations of current data formats in the domain such as MARC[2] and present the process of generating a linked dataset from the existing library cataloguing data.

8.2.2 The Challenges

Despite many efforts to make cultural heritage data open, there are still several challenges that prevent digital cultural heritage archives from being collected and curated using a bottom-up approach, i.e. directly by community groups. Among these challenges are data heterogeneity and the wide range of computer literacy across the cultural heritage community.

Data Heterogeneity

Data formats between collections and tools for digitising heritage data are not consistent. For example, some groups may choose to use common multipurpose tools such as spreadsheets to maintain their archives while others may use pre-existing genealogy software or a relational database. Larger organisations such as national institutions may choose any other format to meet their specific requirements such as key-value data stores for performance or RDF triple stores for integration and reusability. Due to the wide-ranging and different types of data formats, it is not trivial for digitalised cultural heritage to be reused and integrated with each other. Therefore, many such digital archives can only be exploited separately, meaning that connecting local cultural heritage with national archives cannot be done easily. The integration of knowledge from different digital archives, if possible, is done only by human beings. To automate this process, i.e. to integrate/contextualise contents from different archives, it is important to keep data in an open, integrable and reusable format.

---

[2]http://www.loc.gov/marc.

Different Levels of Computer Literacy Among the Cultural Heritage Community

Because community heritage is often being contributed by volunteers and there is a wide range of computer literacy across individuals as well as organisations, it is challenging to provide software platforms that can be used efficiently across individuals and organisations while still allowing data reusability and integration. Thus, it is crucial that software and tools supporting community groups in creating and maintaining their cultural heritage data must not involve technical complexities and, at the same time, be familiar to the targeted users so that training and education can be minimised. However, the current tools and software for creating Linked Datasets and developing Linked Data applications still require in-depth technical knowledge, which is often not available to the broad community within the cultural heritage domain. A promising approach to tackling this problem is to provide an interface between the Linked Data world (tools and standards) and current tools and platforms that ordinary users are familiar with such as standard Content Management Systems (CMS).

8.2.3 The CURIOS Project

Two limitations mentioned in the previous section are some of the main motivations behind the CURIOS[3] platform. The aim of the CURIOS project is to enable community groups to preserve and maintain their digital cultural heritage sustainably by combining the existing open-source software and open data formats. Firstly, the problem of data integration and reusability can be resolved by using Linked Open Data (e.g. OWL, RDF) as the open data standards. Secondly, to assist individuals and groups with a different level of computer literacy to create and maintain their Linked Datasets, a Linked Data adaptor to the existing CMSs was developed.

By combining Linked Data standards and software with Drupal, a popular open-source CMS, CURIOS provides users with limited knowledge of semantic technology a friendly front-end in order to produce Linked Data (and hence to construct the associated knowledge graphs) without requiring a high level of competency in the underlying technologies (e.g. SPARQL, RDF). In CURIOS, the data entered by users are stored in an RDF store while the configuration of how data are presented to users is stored in the Drupal's traditional SQL database. This approach allows the Linked Dataset maintained by CURIOS to be loosely coupled to Drupal, so that it can be reused by different applications and re-purposed in different contexts.

The CURIOS system not only supports the construction of the knowledge graph, but also facilitates services such as semantic searching via the use of SPARQL and the semantic database, configurable presentation and visualisation services to the cultural heritage Linked Datasets for exploiting the graph. It should be noted that

[3] Cultural Repositories and InfOrmation System.

although CURIOS has been deployed in various case studies within the cultural heritage domain, it can still be used as a general-purpose platform which can be applied in other domains.

Within the cultural heritage domain, CURIOS has been used in the following case studies.

Hebridean Connections has been the main CURIOS case study and was carried out in collaboration with historical societies based in the Western Isles of Scotland. Previously, cultural heritage data about the area was collected, archived and presented using a proprietary software. However, there were several limitations with this approach that did not allow the collections to be maintained in a sustainable way. Firstly, data entry had to be done via proprietary software and hence limited the collaborative contribution of volunteers given the uncertain funding sources. Secondly, the archive was kept in a relational database, which made it difficult to be re-used, re-purposed or integrated into other cultural datasets. Based on the original (relational) database schema and suggestions from the historical societies, an ontology for modelling Hebridean Connections archives has been constructed and a subset of this ontology has become the CURIOS upper ontology, the starting point for constructing other knowledge graphs.

Portsoy is another case study about using the CURIOS platform to preserve the cultural heritage of Portsoy, a small fishing village on the North East coast of Scotland. In this case study, we also worked with the local historical society in constructing the ontology modelling their cultural heritage data. The main difference between this study and the Hebridean Connections one is that there was no database from which we could construct an ontology, and hence significant effort for knowledge engineering tasks was required.

CURIOS Mobile is an extension of the CURIOS project that explores how cultural heritage Linked Datasets can be exploited for tourist mobile apps in a rural context [176]. In this project, the Hebridean Connections Linked Dataset was presented on mobile devices. In addition, we investigated how semantic technology could be used to improve tourist experience in rural areas, where the Internet connection is either missing or unreliable. To overcome the connectivity issues, different caching algorithms based on the knowledge graphs have been proposed. A mechanism to rank the records (URIs) based on the level of interest and a narrative generator from RDF triples were also presented.

POWKist is a follow-up project of CURIOS, which focuses on smaller scale collections such as personal diaries, shoe boxes, etc. The main case study in this project is the diaries of Allan Houston, a prisoner of war (POW), during his time in World War II camps. Unlike the main CURIOS project, POWKist investigates how best to visualise cultural heritage data in an "exhibition" format, meaning that only a selection of the datasets are picked, curated and presented to the viewers. In this project, we also explore how the navigation of linked data-based contents can be enhanced to deliver a higher user experience while browsing the collections.

Funeralscapes looks at using CURIOS in a different context, namely, for storing research data and supporting academic collaborative work. In this case study, research data about pre-Christian and Viking ancient burial sites, including text- and media-based materials, are preserved, linked and presented using the CURIOS system. Audio and video materials are hosted by other services such as www.http://soundcloud.com and www.http://youtube.com and presented on a CURIOS Web site via *oEmbed* formats.[4] This approach brings extra flexibility and scalability into the CURIOS system as popular media stores supporting *oEmbed* (e.g. Instagram, Flickr, Youtube) can be used to host media files, a popular means to preserve and present cultural heritage data.

8.2.4 *Constructing the Knowledge Graph*

The knowledge graphs constructed for these case studies were built on top of an *upper ontology*, which specifies the key classes and properties to be used in the extended ontologies. For example, all record types are a subclass of `hc:Subject` and have `dc:title` and `dc:description` (used as in the Dublin Core Schema)[5] to specify the title and description of a record. In addition, the upper ontology specifies other special classes and properties such as the ones holding metadata or the ones used to visualise images and media items.

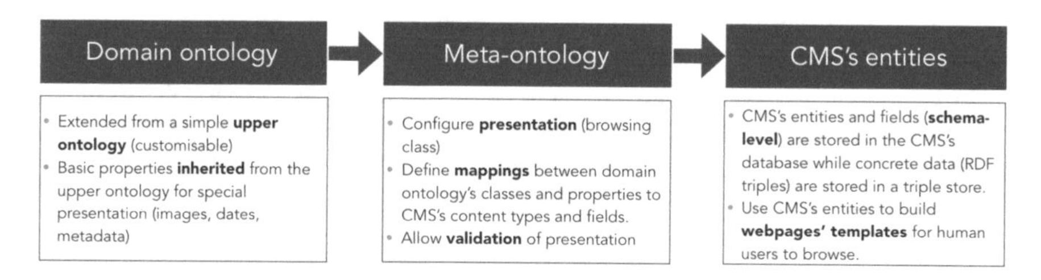

Fig. 8.1 Mapping from a domain ontology to CMS's entities

The whole process of constructing a CURIOS knowledge graph is summarised in Fig. 8.1. Currently, constructing the domain ontology (i.e. an ontology modelling the archive) is the step that consumes most time and efforts in each CURIOS installation. This is due to the fact that the end users often do not have sufficient background on ontologies and modelling techniques to design an ontology that suits their needs best. To assist users in constructing their own domain ontology, we use parts of the Hebridean Connections ontology as the upper ontology, as mentioned earlier. As long as the domain ontology is produced, a *meta-ontology* specifying mappings between the classes and properties of the domain ontology and the entity types and

[4]http://www.oembed.com/.

[5]In practice, the subproperties of these data properties are included in the extended ontology, such as `hc:title` and `hc:description` in the Hebridean Connections ontology.

fields of the CMS are auto-generated. Another important role of the meta-ontology is to allow *validation* of the presentation and data entry. For instance, based on the information of domain/range restrictions in the domain ontology, it is possible to only allow certain types of instances to be linked to each other via a particular object property, e.g. the "child of" relationship can only link a person to another person but not any other type. After having the meta-ontology, the corresponding CMS's entity such as entity types and fields are created, attached and linked to each other.

8.2.5 CURIOS—A Linked Data Adaptor for Content Management Systems

When a knowledge base is created, it is essential to provide means to produce and consume such knowledge, i.e. a writer and a reader. Instead of re-inventing the wheel, we built a Linked Data adaptor to a popular CMS, namely Drupal. As each content management system will have its own data structures, these data structures can be mapped onto corresponding ontological data structures such as classes or data/object properties. For instance in Drupal, there are entity types and fields which can then be mapped onto class and properties and vice versa. Certainly some configurations specifying which ontology classes are mapped to which entity types are needed. Given a domain ontology, we can generate such configurations automatically (see Sect. 8.2.4).

To be able to access Linked Datasets (in an RDF triple format), the database must be a triple store instead of relational databases. This type of back-end database therefore requires a different query language, namely SPARQL [192], instead of SQL. Fortunately, a SPARQL query builder is available as a Drupal module, i.e. SPARQL Views [56], which allows users to generate SPARQL queries in a user-friendly manner. However, SPARQL Views requires some background knowledge on the Semantic Technology to generate the correct SPARQL queries and also does not provide facilities to update the RDF database.

Our system, CURIOS, uses SPARQL Views as a dependent module to read data from an RDF database, or in other words, to generate SPARQL queries and present results. This has a couple of advantages. Firstly, the CURIOS users do not need to concern about generating correct SPARQL queries, and hence do not need in-depth knowledge on Semantic Technologies (e.g. RDF, SPARQL). Secondly, updates on the SPARQL language specification can be dealt with by the developer of the SPARQL Views module instead of CURIOS. To overcome the lack of writer for RDF databases, CURIOS integrates a facility to produce UPDATE queries for SPARQL such as `INSERT`, `DELETE`, etc.[6]

[6]Note that there is currently no `UPDATE` statement in SPARQL. An update can be represented as a combination of a deletion followed by an insertion.

8.2.6 *Presenting and Visualising Cultural Heritage Knowledge Graphs*

Constructing knowledge graphs itself only will not make sense without the final step: disseminating knowledge to the wider public. This step is vital in the Cultural Heritage domain as it would attract public awareness, raise funding opportunities and hence increase the sustainability of cultural heritage projects, especially for community-based projects. In this section, we discuss how cultural heritage can be made more accessible to the wider public. In particular, we describe the presentation and visualisation layers of the CURIOS system.

Generally, in CURIOS the presentational and modelling layers of the knowledge graphs are separated to take advantage of the well-defined semantics and the adaptive and flexible user interface. For example, date values can be entered in different formats and the mapping services of different providers such as Ordnance Survey or Open Street Map can be supported.

Web-Based Presentation of Instances in the Knowledge Graphs

The data held in the CURIOS knowledge graphs are presented as Web pages, as discussed in [177]. Each instance in the knowledge graph has a dedicated Web page that shows information of that specific item such as title, description, data properties and object properties. Figure 8.2 demonstrates how an instance of a CURIOS Knowledge Graph, in this case a residence/croft in the island, is presented. Data properties are

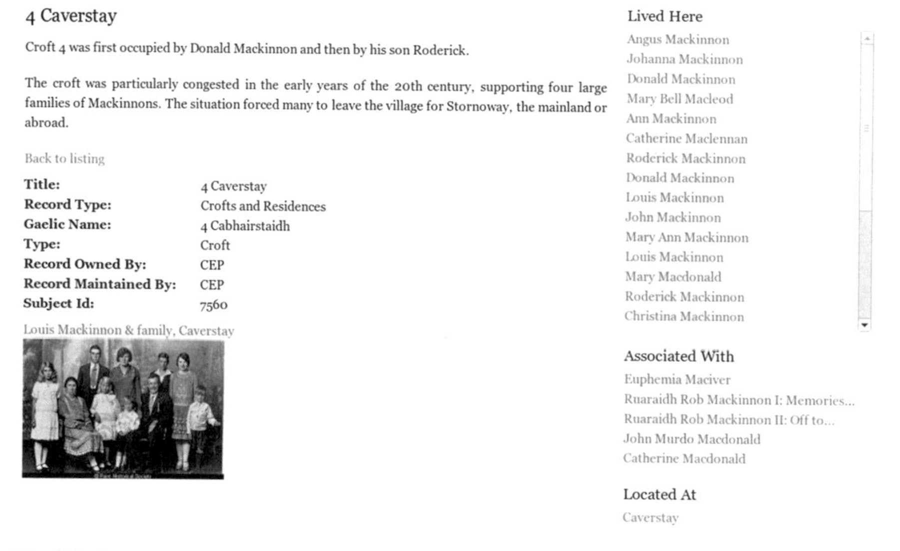

Fig. 8.2 Details of a croft [177]

summarised in an "Info-box", whereas object properties are presented in the form of hyperlinks (on the right) to other instances of Web pages. Some special object properties linking the current instance to instances of media types such as images, audio and video are treated differently. These associative media instances are presented not only as hyperlinks but also as galleries (for images) or suitable embedded media players (for sound and video items). Aggregated presentation of instances is also supported: for example, it is possible to present a collection of the knowledge graph's instances in a table or on a map, as illustrated in Fig. 8.3b, c.

Presenting Data Under Inconsistency and Vagueness

A difficult yet unavoidable challenge when using Linked Data (or any dataset with a well-defined semantic) in the Cultural Heritage is *inconsistency* and *vagueness*. The best example would be the case of temporal data. Figure 8.4 shows some statistics of the date patterns used in the Hebridean Connection corpus [178]. Column 1 describes the categories of date patterns in the corpus. Columns 2 and 3 show the patterns in detail and a representative example of each pattern, respectively. Columns 4 and 5 show the total count (frequency) of each pattern and the total count of the category. The last column indicates whether the pattern can be modelled and presented using the CURIOS system. The first six categories present patterns, which are exact to a specific range of time, e.g. exact to a day, a month, a year up to a century. Note that even the range of time is wide (e.g. up to a century time span), data within these patterns are still interpretable without users' predefined semantics. For example, "Aug 1780" can be interpreted as a date within the range from 01/08/1780 to 31/08/1780 inclusively while "May 12 1780" is clearly interpreted as "12/5/1780". The next four categories describe patterns which are vague (i.e. it is impossible to specify a precise range of time) but can still be modelled and presented given a predefined interpretation. For

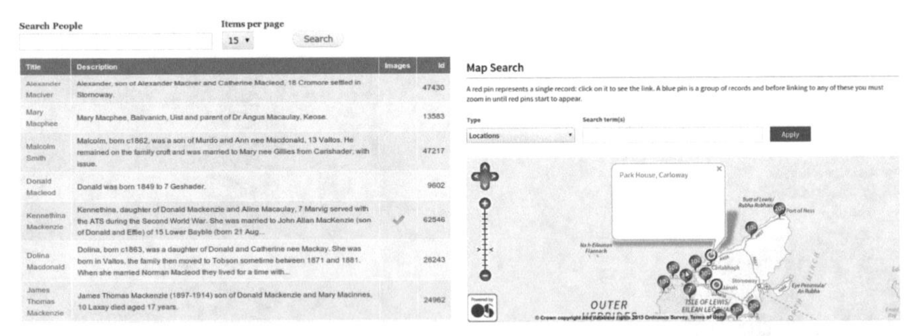

(a) A list of people records [250]

(b) Records with geographical information are aggregated and presented on an Ordnance Survey map

Fig. 8.3 Presenting Linked Data in CURIOS

instance, if the users or modellers agree on a definition of "Winter YYYY" to be the last month of the year "YYYY" and the first 2 months of the following year, it is possible to represent "Winter 1780" as a time period from 01/12/1780 to 28/02/1781. The customisable semantics for temporal data brings flexibility to the CURIOS as different user groups will have different interpretations of a single pattern. As an example, users from Australia might have "Winter YYYY" interpreted in a different way compared to users from Scotland due to the difference in geographical contexts.

As can be seen, only 45 % of the date representation is exact to the date (the first category) and about 10 % (the last category) of date occurrences is uninterpretable as a date. To deal with the remaining cases (making up 45 %), we proposed to use the notion of a date range to model dates in the CURIOS. Some facilities integrate mappings of date values from the presentational layer (e.g. "Winter 1780") to the modelling level (e.g. a date range from 01/12/1780 to 28/02/1781). However, even when a date value is exact, inconsistency remains a problem as the date values are entered into the system in different ways, e.g. some are "y-m-d" while other is "d-m-y". To overcome this, CURIOS provides a user interface for data entry so that date values can be entered in a consistent format (as a date range). For more details on how inexact dates are treated in CURIOS, we refer the readers to [178].

General Class	Pattern	Example	Frequency	Subtotal	Covered
Exact to the day	y-m-d	1780-06-13	12949		
	d-m-y	10/6/45	725		
	d-M-y	12 MAY 1780	272		
	M-d-y	May 12 1780	8		
				13954	yes
Exact to the month	y-m	1780-12	274		
	M-y	Aug 1780	443		
	m-y	03/1780	2		
				719	yes
Exact to the year	y	1978	10825	10825	yes
Exact to the decade	dec	IN 1860'S	1415	1415	yes
Exact to a range of years	y-y	1939-45	242		
	beforey	pre 1918	2		
	aftery	AFT 1890	3		
				247	yes
Exact to the century	cent	20th Century	4	4	yes
Vague within less than a month	mend	Aug/Sept 1972	26	26	yes (using a date range)
Vague within more than a month but less than a year	yend	1978/79	7	7	yes (using a date range)
Vague year	cy	C. 1932	566		
	moddec	early 1950s	86		
				652	yes (using a date range)
Vague around a decade	cdec	c 1950s	2		
	modcent	LATE 1600S	3		
				5	yes (using a date range)
Not directly interpretable as a date	unk	D.I.I.	3069	3069	no
GRAND TOTAL				30923	

Fig. 8.4 Analysis of Date Forms in the Corpus [178]

8.2.7 *Collaborative Construction and Maintenance of Cultural Heritage Knowledge Graphs*

As CURIOS was designed to be a bottom-up approach for collecting and preserving the cultural heritage data, the community groups have been the focus of this project since the beginning. The main difference between the community-level user groups and the institution-level user groups is that the former heavily rely on the contribution of volunteers in terms of time and efforts to create and maintain data. Therefore, CURIOS provides support for collaborative work not only in the construction and maintenance of the knowledge graph but also in the validation of data. This feature is discussed in detail in [250].

Firstly, we use some metadata data properties in the ontology such as "Approved for publication", "Owned by society", "Maintained by society", "Revision notes" to store the information about publication status and revision logs. One might argue as to why the ontology needs to hold such metadata as the facilities for authoring and validating are already available in the CMSs. However, our approach has an advantage of the dataset being loosely coupled to the tool/service that is used to maintain it. For example, the knowledge graph can then be easily exported and edited by other tools such as RDF triple stores or Protege[7] in addition to the Drupal CMS. This is very flexible for bulk update and bulk-import scenarios that often take place in the Cultural Heritage domain.

Secondly, the CURIOS system also employs the roles and permissions feature in the Drupal CMS to design a *flexible user permission scheme* that can be tailored to adapt the organisational structure and policies of different user groups. For example, in the Hebridean Connections case study, permissions of a data creator and a data validator must be mutually exclusive, i.e. a creator cannot validate and publish a record which she has just created. Moreover, a member of a historical society cannot edit or validate a record created by a member of another society. In another case study, Funeralscapes, there is only one group of editors who can create, edit and validate any record.

8.3 Applying Knowledge Graphs in Healthcare

Here we present an application of leveraging knowledge graphs in healthcare application. In this application, we deal with discovering clinical appropriateness in oncology, based on technologies and services of knowledge graphs.

This complicated topic has a big impact on what is called *science/evidence-based medicine*, therefore, on the role of real data and on real knowledge implemented in clinical decision support systems. The final aim is to design technologies and solutions that empower clinicians in their complex decision-making processes. In

[7] http://protege.stanford.edu.

particular, we aim to present the role of knowledge graphs implemented within a clinical decision support system.

The practice of medicine requires the integration of vast and continuously changing information for the prescription of appropriate treatments, therefore, the availability of a powerful tool like dynamic knowledge graph represents a significant advantage when designing decision support systems. Bridging together several information sources is not that simple as it may look like; in particular, when those sources may contain not fully consistent information and whose quality can be compromised because of data entry mistakes, having a knowledge graph approach can be very beneficial to address so many practical issues.

8.3.1 The Problem in Clinical Practice Guidelines

It is well known that diseases are not only major sources of human suffering and one of the most common causes of death worldwide, but they also bring a heavy financial impact to the human society. For example, cancer is one of the most common diseases, besides its tremendous health impact, cancer also bears a staggering financial burden on the world's economy which reached $895 billion, accounting for approximately 1.5 % of the world's GDP in 2008.

In fact, the immense efforts invested to develop cancer treatments have gradually lowered the cancer mortality rates, but have also complicated the process of oncology care. There are now a plethora of cancer treatment options, making it a challenging task to consistently follow the treatment recommendations dictated by *Clinical Practice Guidelines* (CPGs). Interestingly, although deviations from guidelines can have negative results, they are often in fact beneficial. Thus, a central challenge in cancer medical informatics is to identify deviations from CPGs[8] and to assess whether they are medical mistakes or guideline improvements.

The modern medical landscape is characterised by a plethora of different treatment options for even almost indistinguishable clinical statuses. While the development of new treatment modalities is beneficial, it also poses challenges associated with the growing body of evidence on the outcomes of different treatments.

As a consequence of the complexity of treatment possibilities and the presence of widespread variation in medical practice, it has become clear that a large fraction of patients do not in fact receive the best possible care [1, 2]. Deviations from optimal care are abundant in diseases where the treatment efficacy varies as a result of subtle changes in the clinical scenario as well as in cases where clear scientific evidence is not present, as is often seen in cancer [3, 4]. Therefore, an important question in medicine is what leads clinicians to prescribe treatments that do not adhere to the best practice.

[8]CPGs are collective sets of treatment recommendations that attempt to capture the best medical practices for different pathologies [5].

One approach to monitor deviations from standard medical practice is by assessing the adherence to CPGs.

CPGs are promoted as a means to decrease inappropriate practice variation and reduce medical errors [6]. It is generally thought that clinician's adherence to CPG recommendations is the primary means to achieve this goal. High levels of adherence to CPGs may indicate optimal care, whereas low adherence rates may suggest suboptimal treatment. In reality, however, deviation from CPGs often reflects the fact that CPGs cannot be exhaustive; it is not feasible to cover the entire combinatorial space of patient parameters. Deviations from CPG recommendations may thus be beneficial, and it is expected that clinicians will use their personal judgement to contextualise individual patient decisions. In light of the above, previous work identified several barriers to adherence including physician's familiarity with the CPGs, physician's attitudes towards the CPGs, environmental factors, CPG implementation factors and patient-related factors such as preference [7, 8].

Monitoring compliance to CPGs in the clinical setting can be labour intensive. Therefore, in this study we strove to automate the characterisation of adherence to CPGs using Natural Language Processing, data modelling and comparison algorithms. Our vision was to computationally parse *Electronic Health Records* (EHRs) containing both structured and unstructured data to quantify adherence levels, categorise the types of deviations from CPG recommendations, and finally identify the potential rationale for these deviations. We demonstrate our approach using the EHRs of patients diagnosed with adult *Soft-Tissue Sarcoma* (STS). STS is a group of connective-tissue based cancers that account for roughly 1 survival rate of slightly greater than 50, have diverse anatomical origins and can derive from multiple somatic cell types. The variety of histologies results in the presence of multiple drug options and the different anatomical locations offer multiple surgical possibilities. As STSs are rare cancers with numerous treatment options, it is not surprising that prescriptions for patients frequently deviate from CPGs, making STS an ideal use case to evaluate our methodology [10–12].

8.3.2 Preparing the Data and Building the Knowledge Graphs

Data stored within the hospital medical records systems are complex, heterogeneous and stratified over time, and not even necessarily organised by following an intertemporal coherent and effective policy. Clinicians when they have to make decisions they need to face a lot of complexity to explore and integrate a wide range of information. Machines, or decision support systems, they also need a plethora of data that need to be qualified and integrated in a complete, accurate and consistent knowledge base so that useful insight can be extracted out of that. Data curation is therefore extremely important before designing any further services based on top of them and again, the knowledge graph approach is extremely valuable.

Looking at the data, originally they are about clinical tests, patients' medical history and clinical status, guidelines, clinical literature, etc. Specially efforts were made in order to prepare the proper data for building knowledge graphs, which including applying *Natural Language Processing* (NLP) technologies.

In the knowledge graphs, all the data structure is patient-centred and integrates information dealing with standard information such as demographics, clinical status, oncological disease description but also non-oncological information about the patients' eventual comorbidities, etc.

Description of Concepts

The CPGs used in this study were developed by the *Lombardy Oncology Network*, a data sharing network that contains over 50 care premises in Northern Italy. Patient data used in this work was gathered at the Fondazione IRCCS Istituto Nazionale dei tumouri (INT), a network member and thought leader, from November 2006 to November 2012.

The CPGs contained hundreds of clinical cancer presentations (conceptually similar to diagnosis) with their matching recommended treatments. There are multiple recommended treatments for each clinical presentation. A single CPG recommendation was defined as a unique coupling of clinical presentation, recommended treatment and start/end date. The study involved 1484 separate CPG recommendations.

Individual clinical presentations were modelled as a data structure of the following clinical fields: tumour anatomic location, tumour depth (deep/superficial), tumour grade, tumour size, disease status, tumour histological type (liposarcoma, etc.) and surgical status (tumour resectable/not resectable). A clinical presentation can include all or a subset of the fields. This modelling approach is standard for CPGs and is similar to that used by the National Comprehensive Cancer Network [13]. An example of an STS clinical presentation in the Lombardy CPGs is: "Patient with adult soft-tissue sarcoma located in the limb or torso with a deep, high grade, =5cm, localised tumour".

Treatment programmes (TPs) were defined as sequences of medical procedures (treatment elements), for example "Wide surgical excision with adjuvant/neo-adjuvant radiotherapy". A treatment element can contain items such as drug administration, surgery, radiotherapy and transplantation. The Lombardy CPGs contained recommended TPs for each clinical presentation.

Once a physician selected a particular clinical presentation from the CPGs, the matching TPs were presented via the local EHR system. Physicians were entitled to prescribe a TP that was discordant with CPG recommendations (Fig. 8.5). In doing so, they subsequently detailed the contents of their alternatively prescribed TP in free text form. The EHR system recorded this decision, as well as, additional relevant notes provided by the physicians.

Data regarding the treatment was also entered into the EHR system by caregivers during the TP execution. We applied the standard NLP methods on this data to deduce the actual TP that a patient underwent. The extracted TP was compared to the CPG recommended TPs to assess the adherence. The actual TP was considered to deviate if it was discordant with CPG recommendations, regardless of whether the prescription was according to the recommended TPs or not (Fig. 8.5).

Application of the NLP Technique on Electronic Health Records

We applied the NLP techniques on the EHR free text data to computationally retrieve the required information for this study. After Italian to English machine translation, we used the *Unstructured Information Management Architecture* (UIMA) framework to process unstructured information [14]. Our UIMA pipeline included tokenisation, *part-of-speech* (POS) tagging, normalisation using standard terminologies in UMLS (Unified Medical Language System) [15], entity and relationship extraction, semantic analysis, negation and disambiguation reasoning. The resulting structured annotations included drugs, diseases, procedures, symptoms, body regions and tumour characteristics. Relationship extractions were used to infer aspects such as the number of chemotherapy cycles, tumor size, tumour grade and reasons for specific treatment prescription.

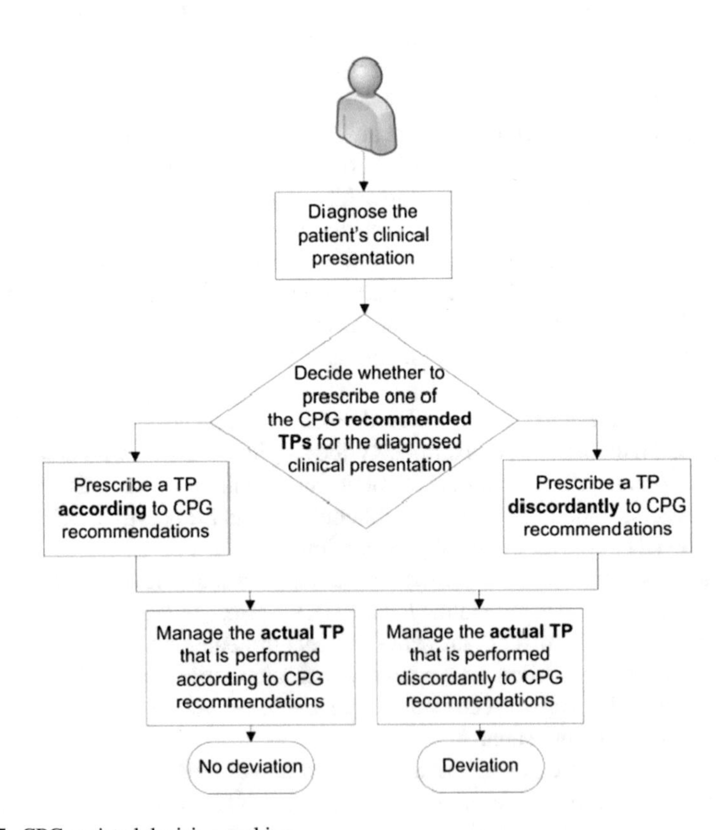

Fig. 8.5 CPG-assisted decision-making

Study Setting, Patient Selection and Data Cleansing

Our patient data encompassed adult STS patients treated at the Fondazione IRCCS Istituto Nazionale dei tumori between November 2006 and November 2012. We acquired 5598 electronic patient discharge letters representing 2699 STS treatment programmes on a total of 2151 different patients. Nine hundred and forty-eight treatment programmes (TPs) with missing data were excluded due to the following reasons: the TPs that were follow-ups, the actual TP was unknown, the TP did not have at least one CPG recommendation due to CPG incompleteness or were clinical studies which were not mentioned in the CPGs. This resulted in 1751 TPs consisting of 1431 patients on whom we performed the analysis presented in this study. Some patients had two or more sequentially prescribed TPs.

8.3.3 Services Based on the Knowledge Graphs

Because of the dynamic feature of the knowledge graphs in this use case, the services should be time-featured, in the sense that the knowledge graph services are dynamic. Based on the decision models, the services can be classified into: discovering deviations, classification of deviations and justification (explanation) for the above results.

Treatment Programme Comparison

The results of the text analytics are structured annotations on the text. The annotations first need to be transformed to a predefined data model to enable advanced analyses. We therefore designed an actual TP model that defines the treatment which was given to patients. The model was designed to enable comparison with the recommended TPs.

To categorise deviations we identified the most similarly recommended TP in the CPGs which was found by assessing the degree of similarity between the recommended and actual TPs. The differences between a deviating actual TP and its most similar recommended TP were classified into different categories.

Extracting Reasons for Deviation

We used the same NLP techniques described above to extract reasons for deviation from CPGs. This was done by identifying relationships between extracted annotations using semantic parsing rules. For example, one can consider the following machine-translated sentence: “In light of extension of illness, the patient’s age and preliminary activity of molecule in this particular histotype, starting chemotherapy with gemcitabine”. By detecting that the conjunction “in light of” connects the two parts of the sentence, we deduced that the first part of the sentence describes reasons for the given treatment, whereas the second part of the sentence ("starting chemotherapy") describes the treatment itself.

Manual Validation

We performed a manual validation of our computational results on a subset of randomly selected TPs. Four human validators were exposed to the entire EHR records and CPGs. Different subsets of the validation dataset were allocated to each reviewer and the results were compiled.

Quantifying Factors That Impact Deviation Frequency

Our computational approach identified deviations in 48.9 % of the actual TPs, meaning that the actual given treatments were found to not fully comply with the CPG recommended TPs.

We next assessed a non-clincial parameter correlation with deviation frequency. Strikingly, 35 % of the TPs prescribed according to CPG recommendations in reality deviated from the CPG recommendations. More expectedly, TPs that were prescribed discordantly to CPG recommendations did in fact deviate in 80 % cases. Gender and age (cutoff set at median age) were not associated with deviation frequency.

Upon analysis of clinical parameters, we observed that all disease and tumour parameters were associated with deviation frequencies, except for tumour location. This analysis portrays an expected trend in which a poorer prognostic status (large, high grade and deep tumours) is linked to substantially higher deviation levels. Indeed, the highest deviation frequency was found in metastatic disease (78 %).

Measuring the Prevalence of Different Deviation Types

TP deviations can be classified using non-mutually exclusive categories. It shows the abundance of deviations that have added or removed treatment elements, differences in chemotherapy drugs, differences in number of chemotherapy cycles and differences in the surgery type.

The most abundant source of deviation was overtreatment, consisting of 39.7 % of all cases in contrast to 12.7 % for missing treatments. Notably, metastatic presentations had no excluded elements despite an overall average of 12.7 % for all TPs. Also prominent was the observation that there was only a 10.3 % deviation rate of type "different chemotherapy drug" for metastatic cases with administered chemotherapy. In general, disease parameters were more strongly associated with chemotherapy differences than surgical differences, with the exception being local/metastatic clinical presentations.

Identification of Potential Reasons for Deviation

NLP parsing identified 1191 potential reasons for deviation among the 857 TPs that we labelled as deviations (average 1.4 per TP). Deviating TP (67.3%) had one to four reasons and 29.7 % had no identified reasons.

The potential reasons for deviation were classified into five categories: cancer status, other clinical, current treatment related, previous treatment related and patient preference related. The reasons for deviation which were based on cancer status represented the majority (59 %) of all deviations.

The reasons for deviation were further classified into lower-level categories. The cancer status category consisted of different tumour and disease progression parameters. Other clinically related reasons included demographics, oncological and non-oncological comorbidities, acute symptoms and overall clinical condition. The previous treatment related reasons include several previous treatments, poor previous response, previously severe side effects, and the presence of residual margins after surgery. The patient preference category included patient treatment requests or refusals. Finally, the current treatment related reasons consisted of anticipated treatment efficacy, impact on the quality of life and newly available clinical evidence. Deviations due to environmental constraints including the lack of personnel or resources were rare and thus not presented.

The largest fraction of deviations appear to result from disease progression or a lack thereof together with the presence of acute symptoms. Interestingly, new medical knowledge was only a small fraction of potential deviation causes.

8.3.4 Contributions to Healthcare Practice

In this section, we introduce a real-world application for a deep understanding of the adherence to CPG recommendations by applying knowledge graphs, used in adult STS clinical studies. The resulting insights from this work significantly contribute to Healthcare Practise on the following aspects:

- **Helping the doctors to understand clinical deviations**. It identifies deviations, classifies them by types and finally proposes reasons that may reflect the physicians' rationale in deviation cases.
- **Helping to understand the decision-making process of physicians**. It offers explanations to the technical directors and managers as to why the deviations are so high. Some reasons could be as given below: (i) the comorbidities are high, (ii) prevalence of malpractice, (iii) CPG is outdated, etc.
- **Improving CPGs**. It can identify cases where the deviations may be beneficial, and to increase the adherence to CPGs when deemed appropriate. In some situations, it can also suggest clinical trials in order to improve the CPGs.

Beyond value in understanding clinical deviations, this current analysis raises multiple observations that may be useful to sarcoma researchers and the decision support community.

Chapter 9
Enterprise Knowledge Graph: Looking into the Future

Jeff Z. Pan, Jose Manuel Gomez-Perez, Guido Vetere, Honghan Wu, Yuting Zhao and Marco Monti

Congratulations! We have covered architecture, technical details and success stories of the Knowledge Graph for large organisations together, in the eight chapters that we have just walked over. In this chapter, we will briefly summarise our journey so far, before providing you some guide on getting started with your first knowledge graph. We will conclude this chapter by some thoughts on "what is next for Knowledge Graph" from a group of experts in the field.

J.Z. Pan (✉)
University of Aberdeen, King's College, Aberdeen AB24 3UE, UK
e-mail: jeff.z.pan@abdn.ac.uk

J.M. Gomez-Perez
Expert System, Prof. Waksman 10, 28036 Madrid, Spain
e-mail: jmgomez@expertsystem.com

G. Vetere
IBM Italia, via Sciangai 53, 00144 Rome, Italy
e-mail: gvetere@it.ibm.com

H. Wu
King's College London, De Crespigny Park, London SE5 8AF, UK
e-mail: honghan.wu@kcl.ac.uk

Y. Zhao · M. Monti
IBM Italia, Circonvallazione Idroscalo, 20090 Milan, Italy
e-mail: yuting.zhao@it.ibm.com

M. Monti
e-mail: marco.monti@it.ibm.com

J.Z. Pan et al. (eds.), *Exploiting Linked Data and Knowledge Graphs in Large Organizations*, DOI 10.1007/978-3-319-45654-6_9

9.1 Conclusion

Whether we notice it or not, Knowledge Graph is changing our ways of accessing information and knowledge: answers to implicit questions behind your searches surface directly from millions of relevant documents when you are googling (Sect. 3.4.3); you are empowered better than ever with the effective access to your colleagues' expertise in problem solving (Sect. 1.3 on p. 5). As evidenced in IBM's cognitive computing system—Watson (Sect. 7.2), Knowledge Graph techniques are exciting new approaches to deal with big data challenges faced by many large organisations. To provide necessary technical details for prospective practitioners in these organisations or other interested readers, this book introduces the core techniques to build enterprise knowledge graphs, to understand and consume them in their typical applications.

Same as taking up any technical task, the necessary background knowledge (Chap. 2) is required for Knowledge Graph solutions, including the RDF data model, OWL ontology languages and Linked Data. In addition to the background, more importantly, we believe a clear bird's-eye view is a must-have to the construction of a successful enterprise knowledge graph and the maintenance of a healthy ecosystem around it.

Chapter 3 presented an abstract reference architecture for creating, maintaining, understanding and exploiting knowledge graphs. The first layer in this architecture deals with knowledge acquisition and integration, wherein we discussed methodologies for defining, capturing and converting the knowledge that is relevant to your organisation. The second layer deals with options for storing and accessing the knowledge graph. Finally, the third layer deals with generic ways to understand and consume the knowledge graph, e.g. by enhanced search, summarisation of graphs and entities as well as question answering. This chapter provided an overview to understand the technical environment in which knowledge graphs exist and provided context in order to understand components and systems described in more detail in Chaps. 4–7 of this book.

The knowledge graph construction involves lifting structured data and compiling knowledge from natural language texts and/or various heterogeneous data sources. To tackle the technical challenges in deriving semantics from the *mess*, an engineering process of *Knowledge Construction and Maintenance Lifecycle* (Sect. 4.1) is proposed to split the task into interlinked subtasks, each of which requires different skill sets and the whole process is supposed to be run iteratively until a satisfactory result is achieved. Two types of construction techniques are introduced: Chap. 4 focuses on the semi-automated ways (i.e. competency question-based ontology authoring and Helix-powered data linkage and integration) and Chap. 5 introduces automated approaches (i.e. scenario-driven entity recognition and Bayesian network based schema learning).

Once you have your Knowledge Graph built, the first question you would probably ask is "Okay, show time! What is the coolest thing I can get from this?" Google's slogan of "things, not strings" points out the most significant feature of Knowledge

Graph applications in a very concise and intuitive way. By introducing the knowledge card to its Web search results, Google made a revolutionary leap in bringing answers directly to search results. Technically, knowledge cards are entity summaries that are generated from underlying Knowledge Graphs, which support a quick and intuitive understanding of entities in the *'Graph'*. In addition to entity summary, Knowledge Graph brings in new possibilities and opportunities for enterprises to make their data more understandable, linkable and reusable across different stakeholders. Chapter 6 has introduced Knowledge Graph understanding techniques from entity level, conceptual level (summarisations) and data source level.

Having learnt that Knowledge Graph makes your enterprise data much easier for access and understanding, it is time to look into how it can further help your enterprise solutions. Methods for organising, finding and selecting relevant information, beyond the capabilities of classic Information Retrieval, are always active topics of research and development. Question answering is one of the most promising among such methods. Knowledge Graphs play a key role in question answering. On the one hand, they are the natural encoding for structured knowledge extracted from texts, databases or other sources, making it available for efficient queries. On the other hand, they provide support for processing textual data. Chapter 7 starts with the introduction of tasks of question answering over text documents. It then describes how question answering can be done over knowledge graphs. In Sect. 7.3 the chapter puts special focus on how knowledge graph approaches helped IBM Watson, which played and won a two-game match against Ken Jennings and Brad Rutter, the biggest all-time money winners in the *Jeopardy!* game history. The rest of this chapter shows how knowledge graphs can improve the performances of a well-studied question-answering framework—UIMA.

Finally (Chap. 8), this book presents successful use cases, in which knowledge graphs reveal their power in depicting and in making even complex and implicit knowledge accessible. In particular, a use case in the healthcare industry is presented together with details on the clinical decision-making processes leveraging the represented knowledge. Innovative analyses, like the gap analysis, designed to identify clinical deviations in respect to the hospital clinical guidelines, are now possible thanks to the adoption of reasoners leveraging the knowledge graph content. The IBM Watson for Healthcare solution empowers clinicians in the assessment of the level of appropriateness of the delivered treatment programmes so as to support the continuous innovation in health care. In this chapter, we aim to show how knowledge graphs represent not just very powerful tools for knowledge representation and exploitation but also new cognitive tools that enhance human–machine interaction and decision-making outcomes.

9.2 Get Started with Knowledge Graphs

As you have learnt, knowledge graphs may be used in many different scenarios, from publishing existing databases in a new powerful form, to extracting and integrating knowledge from many different places; from serving small static datasets, to handling complex information management workflows on huge distributed systems. When designing KG solutions, enterprise architects have to carefully go through the analysis of functional and nonfunctional aspects, as for every information system. Relevant aspects include:

- Provenance: how many different organisations (or branches) supply and maintain the information sources;
- Diversity: how many different data formats, conceptual models and services are to be integrated;
- Dynamics: how often data sources get updated;
- Inquiry: what kind of Information Retrieval should be supported.

As shown in Sect. 4.1, building and maintaining a KG is a cycle of specification, modelling, lifting, publishing and curation. To implement this cycle, architectural criteria are in fact similar to those related to information integration systems, although KGs do not embed or imply any specific integration model. As in any project, in fact, KG system designers should identify the right stakeholders' priorities and trade nonessential features for key functions. For instance, supporting powerful KG query answering on expressive ontologies demands either computational resources at query time, or complex reasoning (and information ingestion) at update time, or approximation. Depending on how the KG is used, one may opt for pushing reasoning (and demanding resources) at query or at update time, or relaxing system's completeness.

When approaching the construction of KGs, useful questions may therefore be as follows:

- How much and how often is the KG going to be queried? How are the queries distributed over time?
- How much and how often is the KG is going to be updated? How are updates distributed over time?
- What kind of query answering tool do KG applications need? How many different queries are issued? What is the level of accuracy requested for?

An appropriate solution depends on how the answers to such kind of basic questions are taken into account. In sum, like any complex information system, the art of building KGs is in balancing (possibly conflicting) requirements. We are confident that previous chapters have provided useful insights to drive this analysis. What follows is a recipe that you can use as a reference to get started with a simple KG implementation.

9.2.1 A Small but Powerful Knowledge Graph

Assume that you have few relatively small datasets to integrate and share within your organisation, and you decide to serve this data through a knowledge graph. Your company wants the new information service to be simple to build and maintain, and yet as powerful as a 'cognitive' system [222]. Plus, your stakeholders want to minimise the impact of the new KG on normal information workflows of the branches which provide datasets. You analyse that a suitable solution, in this case, is to materialise data as an RDF dataset, serve it by means of an RDF store which supports SPARQL and provides reasoning capabilities (see Sect. 3.3.2).

Here is how to proceed:

1. Identify and catalogue your data sources (maybe spreadsheet foils or tables—more likely views—in one or more databases) and make sure you have the means and permissions to extract data from that source.
2. Analyse the entities these data are about. Do you already have a schema (conceptual model) for them? If not, provide it on your own, possibly using an existing (shared) ontology (see Sect. 3.2.1). If yes, consider transforming this legacy schema into an ontology. You may also consider to improve the quality of the legacy conceptual schema: implementing knowledge graphs is a good opportunity to do that. In any case, provide an ontology as a specification of your conceptual model. Make sure that your ontology is satisfiable (i.e. it does not have wrong axioms) using your preferred reasoner. Consider that the ontology may also serve at run-time to support query answering, so its lifecycle must be carefully handled.
3. Focus on the way entities are identified in your dataset, and design suitable identifiers (URIs) for them. Maybe your dataset relies on automatically generated numeric identifiers, or maybe there is a set of data fields which provide unique (composite) access keys, at least for some of the entities in your data. Prefer human-readable URIs whenever it is possible, but beware of name clashes. A good strategy may be to add alphanumeric identifiers (e.g. SSN) to names (e.g. initial and family name).
4. Use your preferred tool (e.g. a scripting/programming language or a mapping tool) to design and implement procedures to extract data for your sources, transform and encode it into RDF (ETL). Keep the ETL module in your code repository, it will be one of the most important modules in your development environment. Consider that your RDF will be generated against your ontology, so make sure that the ontology is stable enough before drawing mappings, regardless of how powerful and easy is your mapping tool: changes in the ontology may require deep revisions of the mapping logic. See also discussions in Sect. 4.2 about data linking.
5. Check for consistency of your generated RDF dataset (possibly, a suitable fragment), using your preferred OWL reasoner. If your dataset is not too heavy, you can easily do this by loading both ontology and data in an open-source authoring tool like Protégé.[1] Remember that your knowledge graph data must satisfy all the constraints you set in your ontology. If there are no problems (i.e. no inconsistencies arise), you can proceed with loading the whole dataset in your triple store.

[1] http://protege.stanford.edu.

6. Design test cases by creating a number of queries (from simple to complex) for which you know the result. Run and evaluate test cases, and check result sets against the ones you expect. If your results are sound (every result item is correct) and complete (every result item you expect is returned), your system has passed the test cases. Now you are ready to put your Knowledge Graph online for users and applications, and monitor its concrete usage.
7. Maintain the KG as a normal data warehousing system, by periodically running the ETL process.

Once you are familiar with this basic process, you could start to try the more advanced competency question driven approach presented in Sect. 4.1. In this advanced approach, some of the test cases are automatically generated from your requirements.

9.2.2 Troubleshooting

Errors may occur at any step of the recipe described above: the sooner they are spotted, the easier it is for you to cope with them. Consistency check at step 5 is crucial for determining the sanity of your KG. Failing a consistency check may reveal that your ontology is ill designed, your ETL contains bugs or some of your data sources are unreliable. Notice that an ontology may be satisfiable (i.e. error free) but may still be inadequate with respect to actual data. For instance, it may set cardinality constraints or disjunctions that do not hold in reality. It might also be that your ontology is correct, but your dataset is noncompliant. This may be caused by either mapping errors or source errors. The former may be hard to detect, and yet they are under your control. The latter may be trivial, but require interactions with data owners.

Consistent ontologies and sound datasets may still give rise to unexpected KG results. This may occur due to technical reasons (e.g. errors in data transfers), but may also reveal subtle semantic entanglements. For instance, if you are using expressive ontologies where you have put nonprimitive concepts (i.e. concepts defined in terms of necessary and sufficient conditions), you may happen to spot out instances failing under concepts that your data have not explicitly asserted. Consider, for example, a concept of `Customer`, which is mapped to a specific table, but it is also defined as the equivalent of `has at least 1 Purchase`. A powerful KB may be able to retrieve customers which are not listed in the source table. Not necessarily, of course, you would blame your KG for these results. But if you want to define concepts in terms of necessary and sufficient conditions (which are not mandatory, however), then you have to be extremely cautious. Is any `Buyer` really a `Client`? Are you sure that your stakeholders would agree with this? Are you sure that they are aware of all the logical consequences of their statements?

9.2.3 Variations

Let us consider other common scenarios, to discuss variations they require to the recipe we have presented in Sect. 9.2.1. When data sources are relevant in size and intensively used (e.g. huge production databases), you may opt for leaving data in their original places, instead of periodically materialising an RDF database. In this case, an OBDA approach is recommended (Sect. 3.3.1). This relieves you from managing the ETL process (step 7); still, mappings (step 4) should be carefully designed, because they will be harder to debug (actually, errors will occur at query time).

If the knowledge of your organisation is provided by data services (e.g. Web Services or APIs) instead of databases, then your KG will act as an intelligent, semantically integrated information hub, based on a single enterprisewide conceptualisation [242]. In this case, similarly to OBDA, data access will be virtual by design, and mappings will play a crucial role.

Let us now consider the case in which (1) your KG is a primary information source, i.e. directly maintained and updated, without dependencies on other databases, and (2) queries needed for the knowledge graph only use the raw vocabulary in your graph, without the use of any high level of application-dependent vocabulary. In this case, you may opt for graph database technologies [11] instead of RDF stores. One of the basic differences is that, as in traditional databases, graph databases do not use the conceptual schemas (ontology) at run-time, hence sophisticated data retrieval patterns may require specific coding. However, for applications based on navigation processes, these technologies provide today viable solution models, specifically for huge datasets. Compliance with W3C Semantic Web standards, if required, may be ensured by means of import/export functions. Note that the condition (2) mentioned above suggests that there is no need to use the schema to bridge high-level application-dependent vocabulary used in queries and the raw vocabulary used in your knowledge graph.

9.3 What is Next: Experts' Predictions into the Future of Knowledge Graph

While finalising the book, the editors of this book had some brief communications with experts in the field on the future visions (Sect. 9.3.1), foreseeable obstacles (Sect. 9.3.2) and suggestions on next steps (Sect. 9.3.3) of Knowledge Graph.

Let us hear what they have to say.

Disclaimer: The views expressed here are solely those of the experts in their private capacities and do not in any way represent the views of the editors.

9.3.1 Future Visions

Editors (of this book): *What is your vision of Knowledge Graph (KG) in enterprise?*

Denny Vrandečić (Ontologist, Google): *Triples are the ultimate basic representation for any form of symbolic knowledge*: everything that can be represented at all in a symbolic knowledge system can be represented in triples. A giant set of such triples, a large graph, can bring together billions of triples from different types of database and knowledge base systems. It can bring together ontological and lexical knowledge, relations between entities, weak and strong taxonomies and much more.

Junlan Feng (Director of Big Data Analytics Lab, China Mobile Research): It had been discussed for years whether taxonomies, ontologies, Semantic Web or knowledge graph would be part of the solution for enterprise information management. *Today, It is not a question anymore.* Internet corporations have proved the power what Knowledge Graph can bring to the business. Traditional enterprises have also started to transform their information system with ontologies and knowledge graphs. For coming years, I believe businesses powered by these technologies will be heating up. This book provides a thorough guide in this line for practisers, researchers and students.

Editors: *As Chris Welty mentioned in his foreword of this book, most information companies, such as Bloomberg, NY Times, Microsoft, Facebook, Twitter and many more, have significant knowledge graphs. What are the implications of this phenomenon?*

Peter Mika (Director of Semantic Search, Yahoo Lab): Companies increasingly realise the value of integrating their structured data assets into unified knowledge graphs in order to exploit the combined knowledge of the enterprise. Compared to previous approaches, knowledge graphs represent a more natural view of data as a connected graph instead of looking at data in multiple tables. As with all data integration technologies, Knowledge Graph also forces companies to unify their view of the world in the form of shared schemas or ontologies and to come up with shared systems of entity identifiers, which are very valuable processes to go through. Further, it is the right time, as two decades of research into the Semantic Web have now produced a set of standards and tools (commercial and open source) that make it easier to implement the concepts introduced in this book.

Sören Auer (Professor of Computer Science, University of Bonn): Investments in digitisation and Big Data technologies in enterprises are an opportunity for knowledge graphs to demonstrate their value for data representation, integration and analytics. With a *paradigm shift* from closed, proprietary IT systems towards more open, standards-based and data-centric information architectures, knowledge graphs can become crystallisation points for enterprise data and a key component for the establishment of data value chains between enterprises.

Juanzi Li (Professor of Computer Science, Tsinghua University): Discovering new knowledge, creating new values and mining new possibilities have become the new cycle for information technology and the service industry. Knowledge is the core competencies for smart products to succeed and an inseparable part for intelligent information processing.

Editors: *What other technologies will complement technologies on constructing and consuming (such as those introduced in this book) enterprise knowledge graphs?*

Alfio M. Gliozzo (Research Manager, IBM Research): My vision is that deep learning technology will change the way we manage knowledge graphs and acquire them from text and other data sources. In fact, deep learning is a very powerful mix of supervised and unsupervised techniques that allow us to deal with several different data sources in a coherent framework, to find associations and sub-symbolic representations and to use them for prediction and classification tasks. I believe that the application of deep learning technology will change the way we interact with knowledge bases, allowing more fluent natural language interface, an easier exploitation for decision support and a convenient framework for inducing them from text and other data sources.

Marco Varone (Founder, President and CTO, Expert System): Knowledge graphs are here to stay: they are the heirs of many other efforts and structures that in the past went nowhere but kept the flame alight while waiting the maturation of many elements. Nobody knows how they will evolve and what they will be in five years but finally we have concrete and solid results on simpler knowledge problems: I think that we will have specialised structures and algorithms for the different use cases and we will always need a lot of human work to create really effective knowledge graphs (automatic learning has strong limitations that cannot be overcome for now) but, with a lot of hard work, we will see good progress that advance the state of the art.

Editors: *What would be the impacts of Knowledge Graph in key application areas such as Healthcare?*

Fabrizio Renzi (Director of Innovation, IBM Italy): Knowledge Graph represents powerful tools for representing and sharing knowledge between human subjects, like in the case of physician-patient interaction, as well as, between humans and computers. Knowledge Graph can complement advanced content analysis by representing the identified concepts discussed in conversations and can support the evolution of the discussions by supporting them with evidence-based knowledge comprehension and inform valuable future decision-making processes.

Richard Dobson (Professor of Medical Bioinformatics, King's College London and The Farr Institute, University College London): There is huge potential for deriving actionable knowledge fròm electronic health records (EHRs) to improve recruitment into trials, tailor treatment for more precision medicine, and streamline operational, financial, and clinical processes, and this is what we are continuously pursuing in NIHR Maudsley BRC. Enterprise knowledge graphs (EKG) sound very promising because of their ability to extract semantically-meaningful knowledge from the unstructured narratives and connect information silos as a graph structure to support multi-mode exploration and exploitation.

9.3.2 Foreseeable Obstacles

Editors: *What are the main obstacles, hindering factors, or difficulties that you see on the road to Knowledge Graph?*

Fabrizio Renzi: Knowledge Graphs are powerful tools, still we need to overcome the obstacle of assessing and validating the reliability of different sources of information, so as to guarantee a valuable comprehension of complex phenomena and mechanisms under investigation.

Denny Vrandečić: *We are really only at the beginning in gathering experience* in how to truly integrate triples from such different sources, in how to maintain such a large dataset, how to ensure the quality, correctness and coverage of such graphs, how to process them efficiently, and much more. And all of this is merely the beginning.

Editors: *What are the main challenges in the social and technical processes behind large-scale KGs?*

Haofen Wang[2] (Assistant Professor of Computing Science, East China University of Science and Technology): There are still several obstacles: For vertical applications, we still lack big ontologies except in a few limited number of domains like healthcare. Also, the current built knowledge graphs are almost facts oriented. They are not capable of capturing dynamic knowledge like process or services. Moreover, the performance of applications heavily depend on the quality and the coverage of KG. Regarding quality, few work has dealt with quality assessment. While the KG cannot be complete, how to handle the incomplete issue and still provide satisfactory effects have not been fully investigated.

Aldo Gangemi (Professor of Computing Science, Université Paris 13): As KGs become widespread, the situation is growing more complex and socially impacting than the classical knowledge engineering community might imagine. Besides technical and scientific problems, there is a socio-political one, due to the fact that the largest KGs are controlled by companies, which typically do not attribute the prove-

[2]Will soon take the role of CTO of Shenzhen Gowild Robotics.

nance of the knowledge in a KG fragment. This should be a worry to all developers and users, and appropriate provenance and trust patterns should be enforced or recommended, be them formally expressed or not.

Juanzi Li: Many challenges have emerged when constructing industry level knowledge graphs. Traditional methods for knowledge graph construction is no longer suitable for today's big data environment, therefore new theories and techniques are required to support the transition to big data, big knowledge and our goal of data to knowledge, to action.

Denny Vrandečić: The current trends in Machine Learning require us to think how to integrate symbolic knowledge graphs with Machine Learning models, and how to extend symbolic knowledge to answer more and more questions and to understand the world. Today, we are still far away from such a goal.

Editors: *In this book, we recommend and present our results on well-understood knowledge representation and data management methods, which come from a long research tradition; is there a concrete risk that KG will miss some relevant finding of this research?*

Tom Heath (Data and Systems Architect, Arup): Knowledge Graph is not a (completely) new concept. In many ways they are the latest incarnation of master data management. The difference is what we've learned from the Web in the intervening years. Once you've internalised its fundamental principles—of universal identifiers, links, and distributed publication—it becomes very hard to unlearn that world view. Knowledge graphs are now transposing those Web principles into data management practices in the enterprise. In doing so, they are also blurring the boundaries between data that originates inside and outside the organisational boundary.

Aldo Gangemi: The relaxation of both syntactic and semantic aspects of KGs has a big potential, but also exposes research and final benefits to hindrances. Most of the advantages brought in the last 50 years (formal semantics, Web identity and resolvability, approximate separation between schema and data, optimised query languages, non-proprietary formats, a rich inventory of design patterns and reusable ontologies, etc.) risk to be put again into the boiler plate for both academic and industrial communities.

Editors: *We talked about healthcare earlier, what are the key foreseeable obstacles in healthcare knowledge graphs?*

Riccardo Bellazzi (Professor of Industrial and Information Engineering, University of Pavia): Since the variety of knowledge sources in healthcare is great, there might be relevant problems in the design phase as well as in the maintenance of the graphs. The solutions to this problem could be on the one hand to take into account clinical data models (such as HL7RIM) as a conceptual basis to build graphs, and on the other hand to exploit as much as possible relational learning to refine structures.

Richard Dobson: There are many challenges in turning electronic health records (EHRs) into Enterprise Knowledge Graphs (EKG). EHRs are often closed, proprietary and there are issues with the accuracy, completeness and the unstructured data they contain. The effective EKG realisation in large organisations like hospitals or NHS Trusts highly depends on the readiness of various technical tools (e.g. ontology authoring, NLP and effective graph data mining) from academic research prototypes to industry level products, and the availability of able, willing and qualified engineers (e.g. data scientists, knowledge engineers and graph data managers).

9.3.3 Suggestions on Next Steps

Editors: *Any thoughts about the next steps?*

Haofen Wang: I believe the next steps of KG evolution will seriously consider these (above) issues in order to build more useful applications in agriculture, telecom, finance, healthcare and other sectors.

Peter Mika: A natural next step for Knowledge Graphs is to extend beyond the boundaries of organisations, connecting data assets of companies along business value chains. This process is still at an early stage, and there is a need for trade associations or industry-specific standards organisations to step in, especially when it comes to developing shared entity identifier schemes. Agreements around schemas in particular can also be facilitated by broader efforts such as schema.org, which was brought to life by the need to share information with search engines regarding the content of Web pages, but has since been used in other contexts such as marking up commercial email content (e.g. receipts). The recently introduced extensions mechanism of schema.org allows a loose integration of industry-led efforts into this broader Web ontology, which means that in the future schema.org could serve as an upper-level ontology for domain-specific efforts.

Oscar Corcho (Professor of Computer Science, Universidad Politécnica de Madrid: In the context of open city data, the benefits that a principled approach to the creation of the knowledge graph of a city potentially offer are many: easier data integration and fusion, better data curation processes, reduced data maintenance costs and better insights into the information that the city handles. Amongst the next steps for knowledge graphs in cities, I would emphasise the establishment of public-private data partnerships, creating a rich ecosystem with additional value both for the cities, their citizens and the local businesses.

Aldo Gangemi: My suggestion for the next steps is to start from the most flexible tool we have devised in the last decades, i.e. pattern-based design, including good practices and reusable components at all design levels: languages, syntaxes, semantic motifs, knowledge patterns, query patterns, reasoning patterns, etc.

Alfio M. Gliozzo: I also believe that a deep learning technology will also play a role to acquire and represent inference patterns between relations and frames that will be needed for reasoning and prediction. To my knowledge, this direction has not been explored yet. If successful it will enable the development of a next generation of reasoning algorithms working on hybrid symbolic and sub-symbolic representations.

Editors: Thank you all for sharing with us your insightful thoughts. We hope you folks can join us again for our next Knowledge Graph book(s).

References

1. RDF 1.1 N-Quads. http://www.w3.org/TR/n-quads/
2. RDF 1.1 N-Triples. http://www.w3.org/TR/n-triples/
3. RDF 1.1 XML syntax. http://www.w3.org/TR/rdf-syntax-grammar/
4. RDFa. http://rdfa.info/
5. Abbasi, R., Staab, S., Cimiano, P.: Organizing resources on tagging systems using t-org. In: In Proceedings of Workshop on Bridging the Gap Between Semantic Web and Web 2.0 at ESWC 2007, June 2007. http://www.uni-koblenz.de/~abbasi/publications/Abbasi2007ORO.pdf (2007)
6. Adrian, B., Sauermann, L., Roth-Berghofer, T.: Contag: a semantic tag recommendation system. In: Pellegrini, T., Schaffert, S. (eds.) Proceedings of I-Semantics' 07, pp. 297–304. JUCS, 2007. ISSN0948-6968. http://www.dfki.uni-kl.de/~sauermann/papers/horak+2007a.pdf (2007)
7. Alexander, K., Cyganiak, R., Hausenblas, M., Zhao, J.: Describing Linked Datasets with the VoID Vocabulary. http://www.w3.org/TR/void/ (1999)
8. Allinson, J.: OpenART: open metadata for art research at the tate. Bull. Am. Soc. Inf. Sci. Technol. **38**(3), 43–48 (2012). ISSN 1550-8366. doi:10.1002/bult.2012.1720380311
9. Altman, N.S.: An introduction to kernel and nearest-neighbor nonparametric regression. Am. Stat. **46**(3), 175–185 (1992)
10. An, Y., Mylopoulos, J.: Translating XML web data into ontologies. In: OTM Workshops, pp. 967–976 (2005)
11. Angles, R., Gutierrez, C.: Survey of graph database models. ACM Comput. Surv. **40**(1), 1:1–1:39 (2008). ISSN 0360-0300. doi:10.1145/1322432.1322433
12. Armstrong, T.G., Ponnekanti, V., Borthakur, D., Callaghan, M.: Linkbench: a database benchmark based on the facebook social graph. In: Proceedings of the 2013 International Conference on Management of Data, pp. 1185–1196. ACM (2013)
13. Artale, A., Calvanese, D., Kontchakov, R., Zakharyaschev, M.: The DL-Lite family and relations. J. Artif. Intell. Res. (JAIR) **36**, 1–69 (2009)
14. Auer, S., Bizer, C., Kobilarov, G., Lehmann, J., Cyganiak, R., Ives, Z.: Dbpedia: a nucleus for a web of open data. In: Proceedings of the 6th International the Semantic Web and 2nd Asian Conference on Asian Semantic Web Conference, ISWC'07/ASWC'07, pp. 722–735. Springer, Berlin (2007). ISBN 3-540-76297-3, 978-3-540-76297-3
15. Baader, F., Bienvenu, M., Lutz, C., Wolter, F., et al.: Query and predicate emptiness in description logics. In: Proceedings of the KR2010
16. Baader, F., Calvanese, D., McGuinness, D.L., Nardi, D., Patel-Schneider, P.F. (eds.): The Description Logic Handbook: Theory, Implementation, and Applications. Cambridge University Press (2003). ISBN 0-521-78176-0

J.Z. Pan et al. (eds.), *Exploiting Linked Data and Knowledge Graphs in Large Organizations*, DOI 10.1007/978-3-319-45654-6

17. Baader, F., Horrocks, I., Sattler, U.: Description logics as ontology languages for the semantic web. In: Festschrift in Honor of JÃűrg Siekmann, Lecture Notes in Artificial Intelligence, pp. 228–248. Springer (2003)
18. Baader, F., Nutt, W.: The Description Logic Handbook, pp. 43–95. Cambridge University Press, New York (2003). ISBN 0-521-78176-0
19. Bail, S., Parsia, B., Sattler, U.: Justbench: a framework for owl benchmarking. In: The Semantic Web–ISWC 2010, pp. 32–47. Springer (2010)
20. Barrasa, J.: Modelo para la definición automática de correspondencias semánticas entre ontologías y modelos relacionales. Ph.D. thesis, Facultad de Informatica, Universidad Politecnica de Madrid, Madrid, Spain (2007)
21. Barrasa, J., Corcho, O., Gómez-Pérez, A.: R2O, an extensible and semantically based database-to-ontology mapping language. In: Second Workshop on Semantic Web and Databases (SWDB2004) (2004)
22. Beaver, D.: Presupposition. In: van Benthem, J., ter Meulen, A. (eds.) The Handbook of Logic and Language, pp. 939–1008. Elsevier (1997)
23. Belhajjame, K., Corcho, O., Garijo, D., Zhao, J., Missier, P., Newman, D., Palma, R., Bechhofer, S., GarcÃa-Cuesta, E., Gómez-Pérez, J.M., et al.: Workflow-centric research objects: first class citizens in scholarly discourse
24. Benjamins, V.R.: Near-term prospects for semantic technologies. IEEE Intell. Syst. **23**(1), 76–88 (2008)
25. Berant, J., Chou, A., Frostig, R., Liang, P.: Semantic parsing on freebase from question-answer pairs. In: EMNLP, pp. 1533–1544. ACL (2013). ISBN 978-1-937284-97-8. http://dblp.uni-trier.de/db/conf/emnlp/emnlp2013.html#BerantCFL13
26. Berant, J., Dagan, I., Goldberger, J.: Learning entailment relations by global graph structure optimization. Comput. Linguist. **38**(1), 73–111 (2012). http://dblp.uni-trier.de/db/journals/coling/coling38.html#BerantDG12
27. Berners-Lee, T., Fielding, R., Masinter, L.: Uniform Resource Identifier (URI): Generic Syntax (RFC 3986). http://www.ietf.org/rfc/rfc3986.txt (2005)
28. Berners-Lee, T., Hendler, J., Lassila, O.: The semantic web. Sci. Am. **284**(5), 34–43 (2001)
29. Bernstein, A., Kaufmann, E., Kaiser, C.: Querying the semantic web with ginseng: a guided input natural language search engine. In: 15th Workshop on Information Technologies and Systems, Las Vegas, NV, pp. 112–126 (2005)
30. Bikakis, N., Giannopoulos, G., Dalamagas, T., Sellis, T.: Integrating keywords and semantics on document annotation and search. In: Proceedings of the 2010 International Conference on the Move to Meaningful Internet Systems: Part II, OTM'10, pp. 921–938. Springer, Berlin (2010). ISBN 3-642-16948-1, 978-3-642-16948-9. http://portal.acm.org/citation.cfm?id=1926129.1926157
31. Bizer, C., Cyganiak, R., Heath, T.: How to Publish Linked Data on the Web. Web page, 2007. Revised 2008. http://www4.wiwiss.fu-berlin.de/bizer/pub/LinkedDataTutorial/ (2007). Accessed 01 Jan 2011
32. Bizer, C., Heath, T., Berners-Lee, T.: Linked data-the story so far. Int. J. Seman. Web Inf. Syst. **5**(3), 1–22 (2009)
33. Bizer, C., Lehmann, J., Kobilarov, G., Auer, S., Becker, C., Cyganiak, R., Hellmann, S.: Dbpedia—a crystallization point for the web of data. Web Seman.: Sci. Serv. Agents World Wide Web **7**(3),154–165 (2009). ISSN 15708268. doi:10.1016/j.websem.2009.07.002. http://wierzba.wzks.uj.edu.pl/09_iracki/eventmarket/uploads/event/summary/2/bizer2009dbpedia.pdf
34. Bizer, C., Schultz, A.: The Berlin SPARQL Benchmark. Int. J. Seman. Web Inf. Syst. (IJSWIS) **5**(2), 1–24 (2009)
35. Bizer, K., Volz, J., Gaedke, M.: Silk—a link discovery framework for the web of data. In: 18th International World Wide Web Conference, pp. 559–572 (2009)
36. Bornea, M.A., Dolby, J., Kementsietsidis, A., Srinivas, K., Dantressangle, P., Udrea, O., Bhattacharjee, B.: Building an efficient RDF store over a relational database. In: SIGMOD, pp. 121–132 (2013)

37. Bouma, G., Parmentier, Y. (eds.): Proceedings of the 14th Conference of the European Chapter of the Association for Computational Linguistics, EACL 2014, April 26–30, 2014, Gothenburg, Sweden. The Association for Computer Linguistics (2014). ISBN 978-1-937284-78-7. http://aclweb.org/anthology-new/E/E14/
38. Brachman, R.J., Levesque, H.J. (eds.): Readings in Knowledge Representation. Morgan Kaufmann Publishers Inc., San Francisco (1985). ISBN 093461301X
39. Brickley, D., Miller, L.: FOAF Vocabulary Specification 0.98. http://xmlns.com/foaf/spec/ (2010)
40. Brin, S., Page, L.: The anatomy of a large-scale hypertextual web search engine. In: Proceedings of the Seventh International Conference on World Wide Web 7, WWW7, pp. 107–117. Elsevier Science Publishers B. V., Amsterdam (1998). http://dl.acm.org/citation.cfm?id=297805.297827
41. Broder, A.Z.: On the resemblance and containment of documents. In: SEQUENCES, pp. 21–29. IEEE Computer Society (1997)
42. Bãűhm, C., Lorey, J., Naumann, F.: Creating void descriptions for web-scale data. Web Seman.: Sci. Serv. Agents World Wide Web **9**(3) (2011). ISSN 1570-8268. http://www.websemanticsjournal.org/index.php/ps/article/view/204
43. Cabrio, E., Palmero Aprosio, A., Cojan, J., Magnini, B., Gandon, F., Lavelli, A.: Qakis @ qald-2. In: Proceedings of the ESWC 2012 Workshop Interacting with Linked Data. Heraklion, Greece (2012)
44. Calvanese, D., Giacomo, G.D., Lembo, D., Lenzerini, M., Rosati, R.: DL-Lite: tractable description logics for ontologies. In: Proceedings of the AAAI (2005)
45. Carbone, F., Contreras, J., Hernández, J.Z., Gomez-Perez, J.M.: Open innovation in an enterprise 3.0 framework: three case studies. Expert Syst. Appl. **39**(10), 8929–8939 (2012)
46. Cardoso, J., Hepp, M., Lytras, M.D.: The Semantic Web: Real-World Applications from Industry, vol. 6. Springer Science & Business Media (2007)
47. Charikar, M.S.: Similarity estimation techniques from rounding algorithms. In: STOC, pp. 380–388. ISBN 1-58113-495-9. http://doi.acm.org/10.1145/509907.509965 (2002)
48. Chatzopoulou, G., Eirinaki, M., Polyzotis, N.: Query recommendations for interactive database exploration. In: Proceedings of the SSDBM, pp. 3–18 (2009)
49. Cheng, G., Ge, W., Wu, H., Qu, Y.: Searching semantic web objects based on class hierarchies. In: LDOW (2008)
50. Christen, P.: A survey of indexing techniques for scalable record linkage and deduplication. IEEE Trans. Knowl. Data Eng. **24**(9), 1537–1555 (2012)
51. Chu-Carroll, J., Fan, J., Boguraev, B., Carmel, D., Sheinwald, D., Welty, C.: Finding needles in the haystack: search and candidate generation. IBM J. Res. Dev. **56**(3.4), 1–6 (2012)
52. Cimiano, P.: Ontology Learning and Population from Text-Algorithms. Springer, Evaluation and Applications (2006). ISBN 978-0-387-30632-2
53. Cimiano, P.: Ontology Learning and Population from Text: Algorithms. Evaluation and Applications. Springer, New York (2006)
54. Cimiano, P., Haase, P., Heizmann, J.: Porting natural language interfaces between domains: an experimental user study with the orakel system. In: Proceedings of the 12th International Conference on Intelligent User Interfaces, IUI '07, pp. 180–189. ACM, New York (2007). ISBN 1-59593-481-2. doi:10.1145/1216295.1216330
55. Cimiano, P., Hotho, A., Staab, S.: Learning concept hierarchies from text corpora using formal concept analysis. CoRR, abs/1109.2140 (2011). http://arxiv.org/abs/1109.2140
56. Clark, L.: SPARQL views: a visual SPARQL query builder for Drupal. In: Polleress, A., Chen, H. (eds.) ISWC Posters and Demos, vol. 658 of CEUR Workshop Proceedings. CEUR-WS.org (2010). http://dblp.uni-trier.de/db/conf/semweb/pd2010.html#Clark10
57. Cruz, I.F., Xiao, H., Hsu, F.: An ontology-based framework for xml semantic integration. In: IDEAS '04: Proceedings of the International Database Engineering and Applications Symposium, pp. 217–226. IEEE Computer Society, Washington, DC (2004). ISBN 0-7695-2168-1. doi:10.1109/IDEAS.2004.10

58. Csomai, A., Mihalcea, R.: Linking documents to encyclopedic knowledge. IEEE Intell. Syst. **23**(5), 34–41 (2008). ISSN 1541-1672. doi:10.1109/MIS.2008.86
59. Curtis, J., Matthews, G., Baxter, D.: On the effective use of cyc in a question answering system. In: IJCAI Workshop on Knowledge and Reasoning for Answering Questions, pp. 61–70 (2005)
60. Cyganiak, R., Reynolds, D.: The RDF Data Cube Vocabulary. http://www.w3.org/TR/vocab-data-cube/ (2010)
61. Cyganiak, R., Wood, D., Lanthaler, M.: RDF 1.1 Concepts and Abstract Syntax. http://www.w3.org/TR/rdf11-concepts/ (2014)
62. d'Amato, C., Fanizzi, N., Esposito, F.: Inductive learning for the semantic web: what does it buy? Seman. Web **1**(1,2), 53–59 (2010). ISSN 1570-0844
63. Damljanovic, D., Agatonovic, M., Cunningham, H.: Freya: an interactive way of querying linked data using natural language. In: Proceedings of the 8th International Conference on the Semantic Web, ESWC'11, pp. 125–138. Springer, Berlin (2012). ISBN 978-3-642-25952-4. doi:10.1007/978-3-642-25953-1_11
64. d'Aquin, M., Motta, E.: Extracting relevant questions to an RDF dataset using formal concept analysis. In: Proceedings of the Sixth International Conference on Knowledge Capture, pp. 121–128 (2011)
65. Davis, M., Whistler, K.: Unicode Normalization Forms. http://www.unicode.org/reports/tr15/ (2010)
66. Demter, J., Auer, S., Martin, M., Lehmann, J.: Lodstats—an extensible framework for high-performance dataset analytics. In: Proceedings of the EKAW 2012, Lecture Notes in Computer Science (LNCS), vol. 7603. Springer (2012)
67. Diab, M.T., Moschitti, A., Pighin, D.: Semantic role labeling systems for Arabic using Kernel methods. In: ACL 2008, Proceedings of the 46th Annual Meeting of the Association for Computational Linguistics, June 15–20, 2008, Columbus, Ohio, USA, pp. 798–806 (2008). http://www.aclweb.org/anthology/P08-1091
68. Doerr, M.: The CIDOC conceptual reference module: an ontological approach to semantic interoperability of metadata. AI Mag. **24**(3), 75 (2003)
69. Dong, X., Gabrilovich, E., Heitz, G., Horn, W., Lao, N., Murphy, K., Strohmann, T., Sun, S., Zhang, W.: Knowledge vault: a web-scale approach to probabilistic knowledge fusion. In: Proceedings of the 20th ACM SIGKDD International Conference on Knowledge Discovery and Data Mining, pp. 601–610. ACM (2014)
70. Duan, S., Fokoue, A., Hassanzadeh, O., Kementsietsidis, A., Srinivas, K., Ward, M.J.: Instance-based matching of large ontologies using locality-sensitive hashing. In: ISWC, pp. 49–64 (2012)
71. Durst, M., Suignard, M.: Internationalized Resource Identifiers (IRIs). http://www.ietf.org/rfc/rfc3987.txt (2005)
72. Dzbor, M., Motta, E., Gomez, J.M., Buil, C., Dellschaft, K., Görlitz, O., Lewen, H.: D4.1.1 analysis of user needs, behaviours & requirements wrt user interfaces for ontology engineering. Technical report (2006)
73. Fanizzi, N., D'Amato, C., Esposito, F.: Machine Learning Methods for Ontology Mining, pp. 131–153. Wiley (2010). ISBN 9780470588222
74. Fellbaum, C.: WordNet: An Electronic Lexical Database. The MIT Press (1998). ISBN 026206197X
75. Fernandez-Lopez, M., Gomez-Perez, A., Juristo, N.: Methontology: from ontological art towards ontological engineering. In: Proceedings of the AAAI97 Spring Symposium, pp. 33–40, Stanford, USA (1997)
76. Ferrández, I., Izquierdo, R., Ferrández, S., Vicedo, J.L.: Addressing ontology-based question answering with collections of user queries. Inf. Process. Manage. **45**(2), 175–188 (2009). ISSN 0306-4573. doi:10.1016/j.ipm.2008.09.001
77. Ferrucci, D.A.: Introduction to "this is watson". IBM J. Res. Dev. **56**(3.4), 1–1 (2012)
78. Ferrucci, D.A., Brown, E.W., Chu-Carroll, J., Fan, J., Gondek, D., Kalyanpur, A., Lally, A., Murdock, J.W., Nyberg, E., Prager, J.M., Schlaefer, N., Welty, C.A.: Building watson: an overview of the deepqa project. AI Mag. 59–79 (2010)

79. Fleischhacker, D., Vãűlker, J.: Inductive learning of disjointness axioms. In: On the Move to Meaningful Internet Systems: OTM 2011, vol. 7045 of Lecture Notes in Computer Science, pp. 680–697. Springer, Berlin (2011). ISBN 978-3-642-25105-4
80. Fokoue, A., Kershenbaum, A., Ma, L., Schonberg, E., Srinivas, K.: The summary abox: cutting ontologies down to size. In: The Semantic Web-ISWC 2006, pp. 343–356. Springer (2006)
81. Fokoue, A., Meneguzzi, F., Sensoy, M., Pan, J.Z.: Querying linked ontological data through distributed summarization. In: AAAI2012
82. Foxvog, D., Bussler, C.: Ontologizing EDI semantics. In: ER (Workshops), pp. 301–311 (2006)
83. Gaasterland, T., Godfrey, P., Minker, J.: An overview of cooperative answering. J. Intell. Inf. Syst. **1**(2), 123–157 (1992)
84. Galárraga, L.A., Teflioudi, C., Hose, K., Suchanek, F.: AMIE: Association rule mining under incomplete evidence in ontological knowledge bases. In: Proceedings of the 22nd International Conference on World Wide Web, WWW '13, pp. 413–422. International World Wide Web Conferences Steering Committee, Republic and Canton of Geneva, Switzerland (2013). ISBN 978-1-4503-2035-1
85. Gangemi, A., Guarino, N., Masolo, C., Oltramari, A.: Sweetening WORDNET with DOLCE. AI Mag. **24**(3), 13–24 (2003). ISSN 0738-4602
86. Gangemi, A., Navigli, R., Velardi, P.: The OntoWordNet Project: extension and axiomatization of conceptual relations in WordNet. In: CoopIS/DOA/ODBASE (2003)
87. García, R., Celma, O.: Semantic integration and retrieval of multimedia metadata. In: Proceedings of the ISWC 2005 Workshop on Knowledge Markup and Semantic Annotation (Semannot'2005) (2005)
88. Gemmell, J., Schimoler, T., Mobasher, B., Burke, R.: Hybrid tag recommendation for social annotation systems. In: Proceedings of the 19th ACM International Conference on Information and Knowledge Management, CIKM '10, pp. 829–838. ACM, New York (2010). ISBN 978-1-4503-0099-5
89. Ghosh, J.K.: Probabilistic networks and expert systems: exact computational methods for bayesian networks by Robert G. Cowell, A. Philip Dawid, Steffen l. lauritzen, David J. Spiegelhalter. Int. Stat. Rev. **76**(2), 306–307 (2008). ISSN 1751-5823
90. Gloor, P.A.: Swarm Creativity: Competitive Advantage Through Collaborative Innovation Networks. Oxford University Press (2005)
91. Gómez-Pérez, J.M., García-Cuesta, E., Garrido, A., Ruiz, J.E., Zhao, J., Klyne, G.: The Semantic Web—ISWC 2013: 12th International Semantic Web Conference, Sydney, NSW, Australia, October 21-25, 2013, Proceedings, Part II, chapter When History Matters—Assessing Reliability for the Reuse of Scientific Workflows, pp. 81–97. Springer, Berlin (2013). ISBN 978-3-642-41338-4. doi:10.1007/978-3-642-41338-4_6
92. Görlitz, O., Thimm, M., Staab, S.: Splodge: systematic generation of sparql benchmark queries for linked open data. In: Proceedings of the ISWC 2012, pp. 116–132. Springer (2012)
93. Grau, B.C., Stoilos, G.: What to ask to an incomplete semantic web reasoner? In: IJCAI, pp. 2226–2231 (2011)
94. Gruhl, D., Nagarajan, M., Pieper, J., Robson, C., Sheth, A.: Context and domain knowledge enhanced entity spotting in informal text. In: Proceedings of the 8th International Semantic Web Conference, ISWC '09, pp. 260–276. Springer, Berlin (2009). ISBN 978-3-642-04929-3
95. Guarino, N., Welty, C.: An overview of OntoClean. In: Staab, S., Studer, R. (eds.) Handbook on Ontologies, International Handbooks on Information Systems, pp. 201–220. Springer, Berlin (2009). ISBN 978-3-540-70999-2. doi:10.1007/978-3-540-92673-3_9
96. Guide, S.U.: Statistical Data and Metadata Exchange Initiative. http://sdmx.org/wp-content/uploads/2009/02/sdmx-userguide-version2009-1-71.pdf (2009)
97. Guo, Y., Pan, Z., Heflin, J.: Lubm: a benchmark for owl knowledge base systems. Web Seman.: Sci. Serv. Agents. World Wide Web **3**(2), 158–182 (2005)
98. Gutierrez-Cuellar, J., Gomez-Perez, J.M.: Havas 18 lab: a knowledge graph for innovation in the media industry. In: Proceedings of the 13th International Semantic Web Conference (ISWC 2014). Springer (2014)

99. Hahn, U., Schulz, S.: Towards a broad-coverage biomedical ontology based on description logics. In: Pacific Symposium on Biocomputing, pp. 577–588 (2003)
100. Hahn, V.: Turning informal thesauri into formal ontologies: a feasibility study on biomedical knowledge re-use. Comp. Funct. Genomics **4**, 94–97(4) (2003). doi:10.1002/cfg.247. http://www.ingentaconnect.com/content/jws/cfg/2003/00000004/00000001/art00247
101. Hakkarainen, S., Hella, L., Strasunskas, D., Tuxen, S.: A semantic transformation approach for ISO 15926. In: Proceedings of the OIS 2006 First International Workshop on Ontologizing Industrial Standards (2006)
102. Hales, B., Pronovost, P.: The checklist a tool for error management and performance improvement. J. Crit. Care **21**(3), 231–235 (2006). ISSN 08839441. doi:10.1016/j.jcrc.2006.06.002
103. Hassanali, K.-N., Hatzivassiloglou, V.: Automatic detection of tags for political blogs. In: Proceedings of the NAACL HLT 2010 Workshop on Computational Linguistics in a World of Social Media, WSA '10, pages 21–22. Association for Computational Linguistics, Stroudsburg, PA, USA (2010). http://portal.acm.org/citation.cfm?id=1860667.1860678
104. Hassanzadeh, O., Chiang, F., Miller, R.J., Lee, H.C.: Framework for evaluating clustering algorithms in duplicate detection. PVLDB **2**(1), 1282–1293 (2009)
105. Hassanzadeh, O., Pu, K.Q., Yeganeh, S.H., Miller, R.J., Hernandez, M., Popa, L., Ho, H.: Discovering linkage points over web data. PVLDB **6**(6), 444–456 (2013)
106. Hassell, J., Aleman-Meza, B., Arpinar, I.B.: Ontology-driven automatic entity disambiguation in unstructured text. In: Proceedings of the 5th International Conference on the Semantic Web, ISWC'06, pp. 44–57. Springer, Berlin (2006). ISBN 3-540-49029-9, 978-3-540-49029-6
107. Hausenblas, M.: The Statistical Core Vocabulary. http://sw.joanneum.at/scovo/schema.html
108. He, H., Garcia, E.A.: Learning from imbalanced data. IEEE Trans. Knowl. Data Eng. **21**(9), 1263–1284 (2009). ISSN 1041-4347
109. Heath, T., Bizer, C.: Linked Data: Evolving the Web into a Global Data Space, vol. 1. Morgan & Claypool (2011). http://linkeddatabook.com/
110. Heino, N., Pan, J.Z.: RDFS reasoning on massively parallel hardware. In: Proceedings of ISWC2012
111. Hellmann, S., Lehmann, J., Auer, S.: Learning of OWL class descriptions on very large knowledge bases. Int. J. Seman. Web Inf. Syst. (IJSWIS) **5**(2), 25–48 (2009)
112. Hepp, M.: Products and services ontologies: a methodology for deriving owl ontologies from industrial categorization standards. Int. J. Seman. Web Inf. Syst. **2**(1), 72–99 (2006)
113. Hepp, M.: Possible ontologies: how reality constrains the development of relevant ontologies. IEEE Internet Comput. **11**(1), 90–96 (2007)
114. Hepp, M.: GoodRelations Ontology. http://purl.org/goodrelations/v1 (2011)
115. Hepp, M., de Brujin, J.: GenTax: a generic methodology for deriving OWL and RDF-S ontologies from hierarchical classifications, thesauri, and inconsistent taxonomies. In: Proceedings of the 4th European Semantic Web Conference (ESWC2007). Springer (2007)
116. Hoffart, J., Suchanek, F., Berberich, K., Weikum, G.: Yago2: a spatially and temporally enhanced knowledge base from wikipedia. Artif. Intell. (2012). https://mpi-inf.mpg.de/yago-naga/yago/publications/aij.pdf
117. Hoffart, J., Yosef, M.A., Bordino, I., Fürstenau, H., Pinkal, M., Spaniol, M., Taneva, B., Thater, S., Weikum, G.: Robust disambiguation of named entities in text. In: Proceedings of the Conference on Empirical Methods in Natural Language Processing, EMNLP '11, pp. 782–792. Association for Computational Linguistics, Stroudsburg, PA, USA (2011). ISBN 978-1-937284-11-4
118. Hogan, A., Harth, A., Passant, A., Decker, S., Polleres, A.: Weaving the pedantic web. In: Linked Data on the Web Workshop (LDOW2010) at WWW'2010 (2010)
119. Holst, T.: Structural analysis of unknown rdf datasets via sparql endpoints. http://www.inf.fu-berlin.de/inst/ag-se/theses/Holst13-RDF-structure.pdf (2013)
120. Horrocks, I., Kutz, O., Sattler, U.: The even more irresistible SROIQ. In: KR 2006 (2006)
121. Horrocks, I., Patel-Schneider, P.: KR and reasoning on the semantic web: OWL. In: Handbook of Semantic Web Technologies, pp. 365–398. Springer, Berlin (2011). ISBN 978-3-540-92912-3

122. Hyland, B., Wood, D.: The joy of data-a cookbook for publishing linked government data on the web. In: Linking Government Data, pp. 3–26. Springer (2011)
123. Hyvönen, E., Viljanen, K., Tuominen, J., Seppälä, K.: Building a national semantic web ontology and ontology service infrastructure—the finnonto approach. In: ESWC, pp. 95–109 (2008)
124. Isaac, A.: SKOS Simple Knowledge Organization System Primer. http://www.w3.org/TR/skos-primer (2009)
125. Ji, Q., Gao, Z., Huang, Z.: Reasoning with noisy semantic data. In: Proceedings of the 8th Extended Semantic Web Conference on the Semanic Web: Research and Applications—Volume Part II, ESWC'11, pp. 497–502. Springer, Berlin (2011). ISBN 978-3-642-21063-1
126. Kalyanpur, A., Boguraev, B., Patwardhan, S., Murdock, J.W., Lally, A., Welty, C., Prager, J.M., Coppola, B., Fokoue-Nkoutche, A., Zhang, L., Pan, Y., Qiu, Z.: Structured data and inference in deepqa. IBM J. Res. Dev. **56**(3), 10 (2012). doi:10.1147/JRD.2012.2188737
127. Kalyanpur, A., Boguraev, B.K., Patwardhan, S., Murdock, J.W., Lally, A., Welty, C., Prager, J.M., Coppola, B., Fokoue-Nkoutche, A., Zhang, L., et al.: Structured data and inference in deepqa. IBM J. Res. Dev. **56**(3.4), 1–10 (2012)
128. Keet, C.M., Fernández-Reyes, F.C., Morales-González, A.: Representing mereotopological relations in OWL ontologies with OntoPartS. In: Simperl, E. et al. (eds.) Proceedings of the 9th Extended Semantic Web Conference (ESWC'12), volume in print of LNCS. Springer (2012). http://www.meteck.org/files/OntoPartSESWC12.pdf
129. Keet, C.M., Khan, M.T., Ghidini, C.: Ontology authoring with forza. In: He, Q., Iyengar, A., Nejdl, W., Pei, J., Rastogi, R. (eds.) CIKM, pp. 569–578. ACM (2013). ISBN 978-1-4503-2263-8. http://dblp.uni-trier.de/db/conf/cikm/cikm2013.html#KeetKG13
130. Kersting, K., De Raedt, L.: Towards combining inductive logic programming with Bayesian networks. In: Inductive Logic Programming, volume 2157 of Lecture Notes in Computer Science, pp. 118–131. Springer, Berlin (2001). ISBN 978-3-540-42538-0. doi:10.1007/3-540-44797-0_10
131. Kim, J.-D., Cohen, K.: Natural language query processing for sparql generation: a prototype system for SNOMED-CT. In: Proceedings of the BioLINK SIG, pp. 32–38 (2013). http://biolinksig.org/proceedings/2013/biolinksig2013_Kim_Cohen.pdf
132. Kimball, R., Caserta, J.: The Data Warehouse ETL Toolkit: Practical Techniques for Extracting. Wiley, Cleanin (2004). ISBN 0764567578
133. Kleb, J., Abecker, A.: Entity reference resolution via spreading activation on rdf-graphs. In: Proceedings of the 7th International Conference on the Semantic Web: Research and Applications—Volume Part I, ESWC'10, pp. 152–166. Springer, Berlin (2010). ISBN 3-642-13485-8, 978-3-642-13485-2
134. Kleinberg, J.M.: Authoritative sources in a hyperlinked environment. J. ACM **46**(5), 604–632 (1999). ISSN 0004-5411. doi:10.1145/324133.324140
135. Kolas, D.: A benchmark for spatial semantic web systems. In: International Workshop on Scalable Semantic Web Knowledge Base Systems (2008)
136. Koller, D., Friedman, N.: Probabilistic Graphical Models: Principles and Techniques—Adaptive Computation and Machine Learning. The MIT Press (2009). ISBN 0262013193, 9780262013192
137. Koller, D., Levy, A., Pfeffer, A.: P-CLASSIC: a tractable probablistic description logic. In: Proceedings of the Fourteenth National Conference on Artificial Intelligence and Ninth Conference on Innovative Applications of Artificial Intelligence, AAAI'97/IAAI'97, pp. 390–397. AAAI Press (1997). ISBN 0-262-51095-2
138. Kotis, K., Vouros, A.: Human-centered ontology engineering: the HCOME methodology. Knowl. Inf. Syst. **10**(1), 109–131 (2006). ISSN 0219-1377
139. Lauser, B., Sini, M.: From agrovoc to the agricultural ontology service/concept server: an owl model for creating ontologies in the agricultural domain. In: DCMI '06: Proceedings of the 2006 International Conference on Dublin Core and Metadata Applications, pp. 76–88. Dublin Core Metadata Initiative (2006). ISBN 970-692-268-7

140. Lehmann, J.: Learning OWL Class Expressions. Ph.D. thesis, University of Leipzig (2010). Ph.D. in Computer Science, supervisors: Prof. Klaus-Peter Fähnrich, Dr. Sören Auer
141. Lehmann, J., Auer, S., Bühmann, L., Tramp, S.: Class expression learning for ontology engineering. J. Web Seman. **9**, 71–81 (2011)
142. Lehmann, J., Hitzler, P.: A refinement operator based learning algorithm for the ALC description logic. In: Proceedings of the 17th International Conference on Inductive Logic Programming, ILP'07, pp. 147–160. Springer, Berlin (2008). ISBN 3-540-78468-3, 978-3-540-78468-5
143. Lehmann, J., Hitzler, P.: Concept learning in description logics using refinement operators. Mach. Learn. **78**(1–2), 203–250 (2010). ISSN 0885-6125
144. Lenzerini, M.: Data integration: a theoretical perspective. In: Proceedings of the Twenty-First ACM SIGMOD-SIGACT-SIGART Symposium on Principles of Database Systems, PODS '02, pp. 233–246. ACM, New York, NY, USA (2002). ISBN 1-58113-507-6. doi:10.1145/543613.543644
145. Li, N., Motta, E.: Evaluations of user-driven ontology summarization. In: Cimiano, P., Pinto, H.S. (eds.) EKAW, volume 6317 of Lecture Notes in Computer Science, pp. 544–553. Springer (2010). ISBN 978-3-642-16437-8. http://dblp.uni-trier.de/db/conf/ekaw/ekaw2010.html#LiM10
146. Liu, K., Fang, B., Zhang, W.: Ontology emergence from folksonomies. In: Proceedings of the 19th ACM International Conference on Information and Knowledge Management, CIKM '10, pp. 1109–1118. ACM, New York, NY, USA (2010). ISBN 978-1-4503-0099-5. doi:10.1145/1871437.1871578
147. Liu, T.-Y.: Learning to Rank for Information Retrieval. Springer (2011)
148. Lopez, V., Fernandez, M., Motta, E., Stieler, N.: Poweraqua: supporting users in querying and exploring the semantic web. In: Semantic Web—Interoperability, Usability, Applicability (2011)
149. Lopez, V., Uren, V., Motta, E., Pasin, M.: Aqualog: an ontology-driven question answering system for organizational semantic intranets. Web Semant. **5**(2), 72–105 (2007). ISSN 1570-8268. doi:10.1016/j.websem.2007.03.003
150. Jarrar, M., Meersman, R.: Ontology Engineering—The DOGMA Approach. In: Advances in Web Semantics, Volume I, LNCS 4891 (2004)
151. Ma, L., Yang, Y., Qiu, Z., Xie, G., Pan, Y., Liu, S.: Towards a complete owl ontology benchmark. In: Sure, Y., Domingue, J. (eds.) The Semantic Web: Research and Applications, volume 4011 of Lecture Notes in Computer Science, pp. 125–139. Springer, Berlin (2006). ISBN 978-3-540-34544-2. doi:10.1007/11762256_12
152. Maala, M.Z., Delteil, A., Azough, A.: A conversion process from flickr tags to rdf descriptions. In: SAW (2007)
153. Maali, F., Cyganiak, R.: Re-using Cool URIs: Entity Reconciliation Against LOD Hubs. Library (2011). http://events.linkeddata.org/ldow2011/papers/ldow2011-paper11-maali.pdf
154. Makela, E., Hyvönen, E., Ruotsalo, T.: How to deal with massively heterogeneous cultural heritage data: lessons learned in culturesampo. Seman. Web **3**(1), 85–109 (2012). ISSN 1570-0844. doi:10.3233/SW-2012-0049
155. Masolo, C., Borgo, S., Gangemi, A., Guarino, N., Oltramari, A.: WonderWeb Deliverable D18 Ontology Library (final). Technical report, IST Project 2001-33052 WonderWeb: Ontology Infrastructure for the Semantic Web (2003)
156. McAfee, A.P.: Enterprise 2.0: the dawn of emergent collaboration. MIT Sloan. Manage. Rev. **47**(3), 21–28 (2006)
157. Mendes, P.N., Jakob, M., García-Silva, A., Bizer, C.: Dbpedia spotlight: shedding light on the web of documents. In: Proceedings of the 7th International Conference on Semantic Systems, I-Semantics '11, pp. 1–8. ACM, New York, NY, USA (2011). ISBN 978-1-4503-0621-8
158. D. C. metadata initiative. Dublin Core metadata element set, version 1.1. http://dublincore.org/documents/dcmi-terms/ (1999)
159. Miles, A., Brickley, D.: SKOS core vocabulary specification. In: W3C Working Draft, World Wide Web Consortium (2005)

160. Milne, D., Witten, I.: An open-source toolkit for mining wikipedia. In: Proceedings of the New Zealand Computer Science Research Student Conference, NZCSRSC, vol. 9 (2009). http://cs.smith.edu/classwiki/images/c/c8/Open_source_mining_wikipedia.pdf
161. Mishne, G.: Autotag: a collaborative approach to automated tag assignment for weblog posts. In: WWW '06: Proceedings of the 15th International Conference on World Wide Web, pp. 953–954. ACM Press, New York, NY, USA (2006). Paper Presented at the Poster Track. http://2006.org/programme/item.php?id=p11
162. Mishra, C., Koudas, N., Zuzarte, C.: Generating targeted queries for database testing. In: Proceedings of SIGMOD (2008)
163. Mitchell, T.M.: Machine Learning. McGraw Hill, New York (1997)
164. Mitchell, T.M., Betteridge, J., Carlson, A., Hruschka, E., Wang, R.: Populating the semantic web by macro-reading internet text. In: Proceedings of the 8th International Semantic Web Conference, ISWC '09, pp. 998–1002. Springer, Berlin (2009). ISBN 978-3-642-04929-3. doi:10.1007/978-3-642-04930-9_66
165. Moschitti, A.: Efficient convolution kernels for dependency and constituent syntactic trees. In: Proceedings of the 17th European Conference on Machine Learning (ECML), pp. 318–329. Springer (2006)
166. Moschitti, A., Chu-Carroll, J., Patwardhan, S., Fan, J., Riccardi, G.: Using syntactic and semantic structural kernels for classifying definition questions in jeopardy! In: Proceedings of the 2011 Conference on Empirical Methods in Natural Language Processing, EMNLP 2011, 27–31 July 2011, John McIntyre Conference Centre, Edinburgh, UK, A meeting of SIGDAT, a Special Interest Group of the ACL, pp. 712–724 (2011). http://www.aclweb.org/anthology/D11-1066
167. Moschitti, A., Patwardhan, S., Welty, C.: Long-distance time-event relation extraction. In: Sixth International Joint Conference on Natural Language Processing, IJCNLP 2013, Nagoya, Japan, October 14–18, 2013, pp. 1330–1338 (2013). http://aclweb.org/anthology/I/I13/I13-1189.pdf
168. Moschitti, A., Quarteroni, S.: Kernels on linguistic structures for answer extraction. In: ACL (2008)
169. Moschitti, A., Quarteroni, S.: Linguistic kernels for answer re-ranking in question answering systems. Inf. Process. Manage. **47**(6), 825–842 (2011). doi:10.1016/j.ipm.2010.06.002
170. Moschitti, A., Quarteroni, S., Basili, R., Manandhar, S.: Exploiting syntactic and shallow semantic kernels for question answer classification. In: Proceedings of ACL-07, pp. 776–783 (2007)
171. Motik, B., Grau, B.C., Horrocks, I., Wu, Z., Fokoue, A., Lutz, C.: Owl 2 web ontology language profiles. W3C Recommendation, 27 October 2009. http://www.w3.org/TR/owl2-profiles/
172. Murdock, J., Kalyanpur, A., Welty, C., Fan, J., Ferrucci, D., Gondek, D., Zhang, L., Kanayama, H.: Typing candidate answers using type coercion. IBM J. Res. Dev. **56**(3.4):7–1 (2012). http://ieeexplore.ieee.org/stamp/stamp.jsp?tp=&arnumber=6177730&isnumber=6177717
173. Neapolitan, R.E.: Learning Bayesian Networks. Prentice-Hall Inc., Upper Saddle River (2003)
174. Nešić, S., Crestani, F., Jazayeri, M., Gašević, D.: Concept-based semantic annotation, indexing and retrieval of office-like document units. In: Adaptivity, Personalization and Fusion of Heterogeneous Information, RIAO '10, pp. 134–135, Paris, France (2010). http://portal.acm.org/citation.cfm?id=1937055.1937088
175. Ngomo, A.-C.N., Auer, S.: Limes—a time-efficient approach for large-scale link discovery on the web of data (2011)
176. Nguyen, H.H., Beel, D.E., Webster, G., Mellish, C., Pan, J.Z., Wallace, C.: CURIOS mobile: linked data exploitation for tourist mobile apps in rural areas. In: Semantic Technology—4th Joint International Conference, JIST 2014, pp. 129–145 (2014). doi:10.1007/978-3-319-15615-6_10
177. Nguyen, H.H., Taylor, S., Webster, G., Jekjantuk, N., Mellish, C., Pan, J.Z., ap Rheinallt, T.: CURIOS: web-based presentation and management of linked datasets. In: Proceedings of the ISWC 2014 Posters & Demonstrations Track a track within the 13th International Semantic

Web Conference, ISWC 2014, Riva del Garda, Italy, October 21, 2014, pp. 249–252 (2014). http://ceur-ws.org/Vol-1272/paper_29.pdf

178. Nguyen, H.H., Taylor, S., Webster, G., Jekjantuk, N., Mellish, C., Pan, J.Z., ap Rheinallt, T., Byrne, K.: A lightweight treatment of inexact dates. In: Semantic Technology—4th Joint International Conference, JIST 2014, Chiang Mai, Thailand, November 9–11, 2014. Revised Selected Papers, pp. 187–193 (2014). doi:10.1007/978-3-319-15615-6_14
179. Niepert, M., Noessner, J., Stuckenschmidt, H.: Log-linear description logics. In: Proceedings of the Twenty-Second International Joint Conference on Artificial Intelligence—Volume Three, IJCAI'11, pp. 2153–2158. AAAI Press (2011). ISBN 978-1-57735-515-1
180. Nikolov, A., dAquin, M.: Identifying Relevant Sources for Data Linking using a Semantic Web Index. Search (2011). http://wtlab.um.ac.ir/parameters/wtlab/filemanager/LD_resources/LDOW2011/ldow2011-paper10.pdf
181. Novalija, I., Mladenic, D., Bradesko, L.: Ontoplus: text-driven ontology extension using ontology content, structure and co-occurrence information. Knowl.-Based Syst. **24**(8), 1261–1276 (2011). http://dblp.uni-trier.de/db/journals/kbs/kbs24.html#NovalijaMB11
182. Oliveira, B., Calado, P., Pinto, H.S.: Automatic tag suggestion based on resource contents. In: Proceedings of the 16th International Conference on Knowledge Engineering: Practice and Patterns, EKAW '08, pp. 255–264. Springer, Berlin (2008). ISBN 978-3-540-87695-3. doi:10.1007/978-3-540-87696-0_23
183. Palmer, S.R., Felsing, M.: A Practical Guide to Feature-Driven Development. Pearson Education (2001)
184. Pan, J.Z.: Description Logics: Reasoning Support for the Semantic Web. Ph.D. thesis, School of Computer Science, The University of Manchester, Oxford Rd, Manchester M13 9PL, UK (2004)
185. Pan, J.Z., Ren, Y., Wu, H., Zhu, M.: Query generation for semantic datasets. In: Proceedings of the Seventh International Conference on Knowledge Capture, pp. 113–116. ACM (2013)
186. Pan, J.Z., Thomas, E., Ren, Y., Taylor, S.: Tractable fuzzy and crisp reasoning in ontology applications. IEEE Comput. Intell, Mag (2012)
187. Parsia, B., Patel-Schneider, P.F.: Owl 2 web ontology language primer. W3C Working Draft, 11 April 2008. http://www.w3.org/TR/2008/WD-owl2-primer-20080411. Accessed 23 June 2008
188. Penela, V., Álvaro, G., Ruiz, C., Córdoba, C., Carbone, F., Castagnone, M., Gómez-Pérez, J.M., Contreras, J.: miKrow: semantic intra-enterprise micro-knowledge management system. In: The Semanic Web: Research and Applications, pp. 154–168. Springer (2011)
189. Phillips, A., Davis, M.: Tags for Identifying Languages. http://tools.ietf.org/html/bcp47 (2009)
190. Prudhommeaux, E., Carothers, G.: RDF 1.1 Turtle: Terse RDF Triple Language. http://www.w3.org/TR/turtle/ (2014)
191. Prud'hommeaux, E., Seaborne, A.: Sparql query language for rdf. W3C Recommendation, 15 January 2008. http://www.w3.org/TR/rdf-sparql-query/
192. Prud'hommeaux, E., Seaborne, A.: SPARQL Query Language for RDF. W3C Recommendation (2008). http://www.w3.org/TR/rdf-sparql-query/
193. Pudota, N., Dattolo, A., Baruzzo, A., Ferrara, F., Tasso, C.: Automatic keyphrase extraction and ontology mining for content-based tag recommendation. Int. J. Intell. Syst. **25**(12), 1158–1186 (2010). ISSN 1098-111X. doi:10.1002/int.20448
194. Quillian, M.R.: Word concepts: a theory and simulation of some basic semantic capabilities. Behav. Sci. **12**(5), 410–430 (1967)
195. Quillian, M.R.: Semantic memory. In: Minsky, M. (ed.) Semantic Information Processing, pp. 227–270. MIT Press (1968)
196. Quinlan, J., Cameron-Jones, R.: Foil: A midterm report. In: Machine Learning: ECML-93, pp. 1–20. Springer (1993)
197. Rahm, E., Bernstein, P.A.: A survey of approaches to automatic schema matching. VLDB J. **10**(4), 334–350 (2001). ISSN 1066-8888. doi:10.1007/s007780100057

198. Rajput, Q., Haider, S.: BNOSA: A Bayesian network and ontology based semantic annotation framework. Web Seman.: Sci. Serv. Agents World Wide Web **9**(2), 99–112 (2011). ISSN 1570-8268. Provenance in the Semantic Web
199. Rector, A., Drummond, N., Horridge, M., Rogers, J., Knublauch, H., Stevens, R., Wang, H., Wroe, C.: Owl pizzas: practical experience of teaching owl-dl: common errors & common patterns. In: Engineering Knowledge in the Age of the Semantic Web. Springer (2004)
200. Ren, Y., Parvizi, A., Mellish, C., Pan, J.Z., van Deemter, K., Stevens, R.: Towards competency question-driven ontology authoring. In: Proceedings of 11th Conference on Extended Semantic Web Conference (ESWC 2014) (2014)
201. Reynolds, D.: The Organization Ontology. http://www.w3.org/TR/vocab-org/ (2014)
202. Rivero, C.R., Schultz, A., Bizer, C., Ruiz, D.: Benchmarking the performance of linked data translation systems. In: LDOW, Citeseer (2012)
203. Rusu, D., Fortuna, B., Mladenic, D.: Automatically annotating text with linked open data. In: Bizer, C., Heath, T., Berners-Lee, T., Hausenblas, M. (eds.) LDOW, volume 813 of *CEUR Workshop Proceedings*. CEUR-WS.org (2011)
204. Sauermann, L.: Pimo-a pim ontology for the semantic desktop (draft). Draft, DFKI (2006).http://www.dfki.uni-kl.de/~sauermann/2006/01-pimo-report/pimOntologyLanguageReport.html
205. Sauermann, L., Cyganiak, R., Ayers, D., Volkel, M.: Cool URIs for the semantic web. Interest Group Note 20080331, W3C. Web page (2008). http://www.w3.org/TR/2008/NOTE-cooluris-20080331/
206. Sawant, U., Chakrabarti, S.: Learning joint query interpretation and response ranking. In: Proceedings of the 22Nd International Conference on World Wide Web, WWW '13, pp. 1099–1110. International World Wide Web Conferences Steering Committee, Republic and Canton of Geneva, Switzerland, 2013. ISBN 978-1-4503-2035-1. http://dl.acm.org/citation.cfm?id=2488388.2488484
207. Schmidt, M., Görlitz, O., Haase, P., Ladwig, G., Schwarte, A., Tran, T.: Fedbench: a benchmark suite for federated semantic data query processing. In: The Semantic Web–ISWC 2011, pp. 585–600. Springer (2011)
208. Severyn, A., Moschitti, A.: Structural relationships for large-scale learning of answer re-ranking. In: Proceedings of the 35th International ACM SIGIR Conference on Research and Development in Information Retrieval, SIGIR '12, pp. 741–750. ACM, New York (2012). ISBN 978-1-4503-1472-5. doi:10.1145/2348283.2348383
209. Severyn, A., Moschitti, A.: Automatic feature engineering for answer selection and extraction. In: EMNLP (2013). http://www.researchgate.net/publication/256708662_Automatic_Feature_Engineering_for_Answer_Selection_and_Extraction/file/72e7e523ac69841c9e.pdf
210. Severyn, A., Nicosia, M., Moschitti, A.: Building structures from classifiers for passage reranking. In: 22nd ACM International Conference on Information and Knowledge Management, CIKM'13, San Francisco, CA, USA, October 27–November 1, 2013, pp. 969–978 (2013). doi:10.1145/2505515.2505688
211. Severyn, A., Nicosia, M., Moschitti, A.: Learning adaptable patterns for passage reranking. CoNLL (2013)
212. Severyn, A., Nicosia, M., Moschitti, A.: Learning semantic textual similarity with structural representations. In: Proceedings of the 51st Annual Meeting of the Association for Computational Linguistics (Volume 2: Short Papers), pp. 714–718. Association for Computational Linguistics, Sofia, Bulgaria, August 2013. http://www.aclweb.org/anthology/P13-2125
213. Shearer, R., Motik, B., Horrocks, I.: HermiT: a highly-efficient OWL reasoner. In: Dolbear, C., Ruttenberg, A., Sattler, U. (eds.) OWLED 2008, volume 432 of *CEUR Workshop Proceedings*. CEUR-WS.org (2008). http://dblp.uni-trier.de/db/conf/owled/owled2008.html#ShearerMH08
214. Shekarpour, S., Ngomo, A.-C.N., Auer, S.: Question answering on interlinked data. In: WWW, pp. 1145–1156 (2013)
215. Siberski, W., Pan, J.Z., Thaden, U.: Querying the semantic web with preferences. In: Proceedings of ISWC2006

216. Sirin, E., Parsia, B., Grau, B., Kalyanpur, A., Katz, Y.: Pellet: a practical OWL-DL reasoner. J. Web Seman. **5**(2), 51–53 (2007). ISSN 1570-8268. http://apps.isiknowledge.com.proxy.library.ucsb.edu:2048/full_record.do?product=WOS&search_mode=GeneralSearch&qid=11&SID=4ClAPHkFckJgGHMNI5N&page=1&doc=2
217. Slutz, D.: Massive stochastic testing of SQL. In: VLDB, pp. 618–622 (1998)
218. Soergel, D., Lauser, B., Liang, A., Fisseha, F., Keizer, J., Katz, S.: Reengineering thesauri for new applications: the agrovoc example. J. Digit. Inf. **4**(4) (2004)
219. Spackman, K.: Managing clinical terminology hierarchies using algorithmic calculation of subsumption: experience with SNOMED-RT. JAMIA (2000)
220. Sporny, M., Kellogg, G., Lanthaler, M.: JSON-LD 1.0. http://www.w3.org/TR/json-ld/ (2014)
221. Steigmillera, A., Liebigb, T., Glimma, B.: Konclude: system description (2013)
222. Stevens, A.: The most popular trends in cognitive computing. Web page (2016). https://www.ibm.com/blogs/watson/2016/05/popular-trends-cognitive-computing/
223. Stojanovic, L., Stojanovic, N., Volz, R.: A reverse engineering approach for migrating data-intensive web sites to the semantic web. In: Proceedings of the Conference on Intelligent Information Processing (2002)
224. Suárez-Figueroa, M.C.: NeOn Methodology for Building Ontology Networks: Specification, Scheduling and Reuse. Ph.D. thesis, Facultad de Informática, Universidad Politécnica de Madrid, Madrid, Spain (2010)
225. Suarez-Figueroa, M.C., Gómez-Pérez, A.: NeOn Methodology for Building Ontology Networks: a Scenario-based Methodology. In: (S3T 2009) (2009)
226. Suárez-Figueroa, M.C., Gómez-Pérez, A., Motta, E., Gangemi, A.: Ontology Engineering in a Networked World. Springer Science & Business Media (2012)
227. Suchanek, F.M., Kasneci, G., Weikum, G.: Yago: a core of semantic knowledge. In: 16th International World Wide Web Conference (WWW). ACM Press, New York (2007)
228. Suchanek, F.M., Sozio, M., Weikum, G.: SOFIE: a self-organizing framework for information extraction. In: Proceedings of the 18th International Conference on World Wide Web, WWW 2009, Madrid, Spain, April 20–24, 2009, pp. 631–640 (2009). doi:10.1145/1526709.1526794
229. Tatu, M., Srikanth, M., D'Silva, T.: Tag recommendations using bookmark content. In: Proceedings of ECML PKDD Discovery Challenge (RSDC08), pp. 96–107 (2008)
230. Tenenbaum, L., Shapira, B., Shoval, P.: Ontology-based classification of news in an electronic newspaper. In: Proceedings of INFOS 2008, Varna, Bulgaria, pp. 89–97 (2008)
231. Thomas, E., Pan, J.Z., Ren, Y.: TrOWL: Tractable OWL 2 Reasoning Infrastructure. In: The Proceedings of the Extended Semantic Web Conference (ESWC2010) (2010)
232. Tran, T., Wang, H., Rudolph, S., Cimiano, P.: Top-k exploration of query candidates for efficient keyword search on graph-shaped (rdf) data. In: ICDE, pp. 405–416 (2009). ISBN 978-0-7695-3545-6. doi:10.1109/ICDE.2009.119
233. Tsarkov, D., Horrocks. I.: FaCT++ Description logic reasoner: system description. In: Proceedings of the International Joint Conference on Automated Reasoning (IJCAR 2006), volume 4130 of *Lecture Notes in Artificial Intelligence*, pp. 292–297. Springer (2006)
234. Tymoshenko, K., Moschitti, A., Severyn, A.: Encoding semantic resources in syntactic structures for passage reranking. In: Proceedings of the 14th Conference of the European Chapter of the Association for Computational Linguistics, pp. 664–672. Association for Computational Linguistics, Gothenburg, Sweden (2014). http://www.aclweb.org/anthology/E14-1070
235. Unger, C., Bühmann, L., Lehmann, J., Ngomo, A.-C.N., Gerber, D., Cimiano, P.: Template-based question answering over RDF data. In: Proceedings of the 21st international conference on World Wide Web, pp. 639–648 (2012). http://jens-lehmann.org/files/2012/tbsl_www.pdf
236. Unger, C., Cimiano, P.: Pythia: compositional meaning construction for ontology-based question answering on the semantic web. In: Proceedings of the 16th International Conference on Natural Language Processing and Information Systems, NLDB'11, pp. 153–160. Springer, Berlin (2011). ISBN 978-3-642-22326-6. http://dl.acm.org/citation.cfm?id=2026011.2026028
237. Uschold, M., Gruninger, M., et al.: Ontologies: principles, methods and applications. Knowl. Eng. Rev. **11**(2), 93–136 (1996)

238. van Assem, M., Gangemi, A., Schreiber, G.: Conversion of WordNet to a standard RDF/OWL representation. In: Proceedings of the Fifth International Conference on Language Resources and Evaluation (LREC'06), Genoa, Italy (2006)
239. van Assem, M., Malaisé, V., Miles, A., Schreiber, G.: A method to convert Thesauri to SKOS. In: The Semantic Web: Research and Applications, pp. 95–109 (2006). doi:10.1007/11762256_10
240. van Assem, M., Menken, M., Schreiber, G., Wielemaker, J.: A method for converting thesauri to RDF/OWL. In: Proceedings of the Third International Semantic Web Conference (ISWC). Springer (2004)
241. Verborgh, R., Hartig, O., De Meester, B., Haesendonck, G., De Vocht, L., Vander Sande, M., Cyganiak, R., Colpaert, P., Mannens, E., Van de Walle, R.: Querying datasets on the web with high availability. In: The Semantic Web–ISWC 2014, pp. 180–196. Springer (2014)
242. Vetere, G., Lenzerini, M.: Models for semantic interoperability in service-oriented architectures. IBM Syst. J. **44**(4), 887–904 (2005). doi:10.1147/sj.444.0887
243. Vila-Suero, D., Villazón-Terrazas, B., Gómez-Pérez, A.: datos.bne.es: a library linked dataset. Seman. Web **4**(3), 307–313 (2013). doi:10.3233/SW-120094
244. Villazón-Terrazas, B., Suárez-Figueroa, M.C., Gómez-Pérez, A.: A pattern-based method for re-engineering non-ontological resources into ontologies. Int. J. Seman. Web Inform. Syst. **6**(4), 27–63 (2010)
245. Villazón-Terrazas, B., Vilches-Blázquez, L.M., Corcho, O., Gómez-Pérez, A.: Methodological guidelines for publishing government linked data. In: Linking Government Data, pp. 27–49. Springer (2011)
246. Villazón-Terrazas, B.M.: A Method for Reusing and Re-engineering Non-ontological Resources for Building Ontologies, vol. 12. IOS Press (2012)
247. Völker, J., Niepert, M.: Statistical schema induction. Seman. Web Res, Appl (2011)
248. Vrandecic, D., Pinto, H.S., Sure, Y., Tempich, C.: The DILIGENT knowledge processes. J. Knowl. Manage. **9**(5), 85–96 (2005). http://www.aifb.uni-karlsruhe.de/WBS/ysu/publications/2005_kmjournal_diligent.pdf
249. Wang, W., Barnaghi, P., Bargiela, A.: Probabilistic topic models for learning terminological ontologies. IEEE Trans. Knowl. Data Eng. **22**, 1028–1040 (2010)
250. Webster, G., Nguyen, H.H., Beel, D.E., Mellish, C., Wallace, C.D., Pan, J.Z.: CURIOS: connecting community heritage through linked data. In: Proceedings of the 18th ACM Conference on Computer Supported Cooperative Work and Social Computing, CSCW 2015, Vancouver, BC, Canada, 14–18 March 2015, pp. 639–648 (2015). doi:10.1145/2675133.2675247
251. Welty, C.A.: OntOWLClean: cleaning OWL ontologies with OWL. In: Bennett, B., Fellbaum, C. (eds.) FOIS, volume 150 of Frontiers in Artificial Intelligence and Applications, pp. 347–359. IOS Press (2006). ISBN 978-1-58603-685-0. http://dblp.uni-trier.de/db/conf/fois/fois2006.html#Welty06
252. Wielinga, B., Schreiber, A.T., Wielemaker, J., Sandberg, J.: From thesaurus to ontology. In: K-CAP '01: Proceedings of the 1st International Conference on Knowledge Capture, pp. 194–201. ACM Press, New York (2001). ISBN 1581133804. doi:10.1145/500737.500767. http://portal.acm.org/citation.cfm?id=500767
253. Wu, H., Qu, Y.: Understanding semantic web entity: concept space based summarization method. J. Southeast Univ. **39**(4), 723–727 (2009)
254. Wu, H., Villazon-Terrazas, B., Pan, J.Z., Gomez-Perez, J.M.: Exploiting semantic web datasets: a graph pattern based approach. In: Proceedings of the Eighth Chinese Semantic Web and Web Science Conference, CCIS, pp. 167–173. Springer (2014)
255. Xia, F., Liu, T.-Y., Wang, J., Zhang, W., Li, H.: Listwise approach to learning to rank: theory and algorithm. In: Proceedings of the 25th International Conference on Machine Learning, ICML '08, pp. 1192–1199. ACM, New York (2008). ISBN 978-1-60558-205-4. doi:10.1145/1390156.1390306

256. Yang, Y., Calmet, J.: OntoBayes: an ontology-driven uncertainty model. In: Proceedings of the International Conference on Computational Intelligence for Modelling, Control and Automation and International Conference on Intelligent Agents, Web Technologies and Internet Commerce Vol-1 (CIMCA-IAWTIC'06)—Volume 01, CIMCA '05, pp. 457–463. IEEE Computer Society, Washington, DC (2005). ISBN 0-7695-2504-0-01
257. Zavitsanos, E., Paliouras, G., Vouros, G.A., Petridis, S.: Learning subsumption hierarchies of ontology concepts from texts. Web Intell. Agent Syst. **8**(1), 37–51 (2010). http://dblp.uni-trier.de/db/journals/wias/wias8.html#ZavitsanosPVP10
258. Zhang, X., Cheng, G., Qu, Y.: Ontology summarization based on rdf sentence graph. In: Williamson, C.L., Zurko, M.E., Patel-Schneider, P.F., Shenoy, P.J. (eds.) WWW, pp. 707–716. ACM (2007). ISBN 978-1-59593-654-7. http://dblp.uni-trier.de/db/conf/www/www2007.html#ZhangCQ07
259. Zhang, Z., Nasraoui, O.: Mining search engine query logs for query recommendation. In: Proceedings of WWW2006
260. Zhao, J., Gomez-Perez, J.M., Belhajjame, K., Klyne, G., Garcia-Cuesta, E., Garrido, A., Hettne, K.M., Roos, M., Roure, D.D., Goble, C.A.: Why workflows break—understanding and combating decay in taverna workflows. In: eScience, pp. 1–9. IEEE Computer Society (2012). ISBN 978-1-4673-4467-8
261. Zhou, L.: Ontology learning: state of the art and open issues. Inf. Technol. Manage. **8**(3), 241–252 (2007). doi:10.1007/s10799-007-0019-5. http://www.springerlink.com/content/j4g2211217k00833/
262. Zhu, M.: DC proposal: ontology learning from noisy linked data. In: Proceedings of the 10th International Conference on the Semantic Web—Volume Part II, ISWC'11, pp. 373–380. Springer, Berlin (2011). ISBN 978-3-642-25092-7
263. Zhu, M., Gao, Z., Pan, J.Z., Zhao, Y., Xu, Y., Quan, Z.: Ontology learning from incomplete semantic web data by BelNet. In: Proceedings of the 2013 IEEE 25th International Conference on Tools with Artificial Intelligence, ICTAI '13, pp. 761–768. IEEE Computer Society, Washington, DC (2013). ISBN 978-1-4799-2971-9

Index

J.Z. Pan et al. (eds.), *Exploiting Linked Data and Knowledge Graphs in Large Organizations*, DOI 10.1007/978-3-319-45654-6

Printed in the United States
By Bookmasters